The
Cellular
Connection

The
Cellular
Connection

A Guide to
Cellular Telephones

Josef Bernard

QUANTUM PUBLISHING, INC.

Editor and Publisher *Paul Mandelstein*
Project Management, Copyediting, and Electronic Composition
    Professional Book Center
Production Assistant *Daniel Bank*
Text design *Editorial Design and Professional Book Center*
Cover design *Naganuma Design*
Pasteup and mechanicals *Paul Koroshetz*
Cartoonist *Tom Durfee*
Illustrator *Wordset/Fred Wakida*

First edition, December 1986
Second edition, 1987

Library of Congress **Cataloging-in-Publication** Data

Bernard, Josef, 1943–
    The cellular connection : a guide to cellular telephones / by
Josef Bernard, — New, rev. ed.
        p.   cm.
    Includes index.
    ISBN 0-930633-08-3
    1. Cellular radio. 2. Telephone.   I. Title.
TK6570.M6B47   1994
384.5'3—dc20   94-5173
                                                          CIP

ISBN 0-930633-08-3

Published by Quantum Publishing, Inc.
P.O. Box 310
Mendocino, California 95640

Contents

Preface

The need to keep in touch—we all have it, whether for business or personal reasons. Now you can keep in touch on the way to work or the shopping center, from a construction site or the golf course, or in a rental car or on your boat.

If you spend time away from your best communication tool—the telephone—a cellular phone can add several business hours to your week. Now you can stay in touch with your office, your customers, or your family, even as you inch along in a traffic jam. And cellular's excellent audio performance ensures clear voice reception.

In today's fast-paced world, the average business manager spends fourteen work weeks per year on the telephone. Salespeople, doctors, and wheeler-dealers report that when they spend time on the road, a cellular phone makes them much more productive. So it's no surprise that cellular has grown to 9 million subscribers and is expected to exceed 25 million before the end of the decade.

Cellular phones are not only for the person who spends time in a car. The advent of portable and transportable phones has created several new applications. Architects, carpenters, or electricians working at construction or repair sites can be in touch with their home offices. Newspaper reporters and other journalists on assignment can transmit their stories to headquarters immediately or send written copy from a portable computer via their portable phone.

And let's not forget about the family—their safety, pleasure, and convenience can be greatly increased by using a cellular phone, both in the car and away from it. Driving to the movies, taking a bus to a sporting event, or just walking to the corner store, instant communications can be at their fingertips.

Although using a cellular phone might be as easy as pushing a few buttons, it is still a large investment, and there are several things you should understand when choosing your phone, and the service you use with it. Everything you need to know is explained here in an easy-to-understand manner.

You will learn how the cellular system works; what are the purposes and advantages of each feature; how to choose a telephone and telephone carrier; how to install and operate your phone; what to do when traveling outside your home area; how to tap into other communciation systems; and what's coming in the future. Also included are informative illustrations, photographs, and a glossary of terms for easy reference.

Welcome to the world of cellular telephones—communications on the move.

1

An Introduction to Cellular Phone Systems

The first mobile communications system began in 1921 when the Detroit Police Department installed two-way radios in its cars. The equipment was bulky, awkward, and a drain on automobile batteries, but it proved its worth. Soon police and fire departments throughout the country were installing two-way radio equipment in their fleets.

Eventually someone realized that private individuals could also benefit from being able to talk with the outside world from their cars, and the first commercial mobile telephone service was instituted in St. Louis, Missouri, in 1946. Early mobile phone service was more like using a radio than a phone. You spoke into a microphone or special handset, and the voice from the other end of the connection usually came through a loudspeaker. You couldn't dial a call; instead, the mobile operator at the other end of your radio link established the connection for you. It could take some time before you got through, since only a few radiotelephone channels were available in any city, and you frequently had to wait for a free one.

When you finally did get through, your conversation was clumsy, at best. Because of the nature of the equipment, you and the party you were speaking with had to take turns talking. If you tried to say something while the other party was speaking, they wouldn't be able to hear you and you wouldn't be able to interrupt them. You had to wait until they decided that they had their say and gave you the go-ahead to speak—a very frustrating experience, as you may well imagine.

Still, it was worth putting up with such hardships to be able to accomplish needed business while on the road or just to enjoy the luxury of chatting with friends as you motored from place to place. Your car, with its impressive buggy-whip antenna, marked you as a very important person. (The concept of mobile phones as status symbols was illustrated by the story of the Hollywood executive who, while speaking to an associate from his car, asked him to hold on for a moment . . . the executive's *other* phone was ringing.)

Improvements in the design of electronic equipment soon made mobile telephones easier to use. In 1948, the first automated mobile dialing system was demonstrated, although it was not used commercially until fifteen years later. Now you didn't need the assistance of the mobile operator—you could place your calls yourself. And you could even dispense with push-to-talk operation and converse almost normally. You still got plenty of busy signals, though, waiting for a channel to open up in a congested area.

In 1969 the Improved Mobile Telephone System (IMTS) was introduced, although it was not much of an improvement over what had existed before. The number of channels was still limited, and in some areas there were waiting lists for mobile phone installations. The range of a particular system was also limited to a radius of between twenty and twenty-five miles from the centrally located transmit-

"Would you please hold for a moment? My other phone is ringing…"

ter, and interference from other phone systems was a common problem. Even so, the demand for mobile phone service was greater than could be met.

In the late 1960s and the 1970s there was a growing awareness of just how inadequate the existing mobile telephone service was, and a search was instituted for a better way. A proposal was made at the end of 1971 for a type of service called *cellular* (the concept for which had existed at least as far back as 1947), and in 1978 a trial cellular service began in Chicago, serving about 2000 customers. Within a year and a half, AT&T had created a subsidiary called Advanced Mobile Phone Service, Inc., to develop and market cellular telephone service nationwide, and in October 1983 the first cellular phone service was inaugurated in Chicago.

THE CELLULAR METHOD

Until the advent of cellular phones, radiotelephone systems—even the Improved Mobile Telephone System (IMTS) mentioned above—worked pretty much the same way. In your car you had a radio transmitter and receiver, and at a point central to your service area was another, more powerful transmitter and receiver operated by the telephone company to which you subscribed. This area transmitter/receiver had the facilities to connect you into the regular telephone lines, through which your conversations with the rest of the nonmobile world took place.

An area or city was served by a single transmitter/receiver location. The IMTS transmitter had a range of twenty to twenty-five miles using a power of perhaps 250 watts. As you got farther away from the central antenna location, signals—both to and from the central site—became weaker and noisier, often making it difficult to maintain a conversation. In addition, if you were in a region with a number of mobile phone services, there was a good chance your conversation might be interrupted by interference from other mobile users.

You were generally restricted to using just the service you subscribed to, and, if you ventured outside your local area, your expensive mobile phone installation became about as useful to you as a fifth wheel.

But cellular service operates in a completely different fashion. Instead of having one central, high-powered transmitter to cover an entire region, the cellular system divides that region into a number of small cells just a few miles across. (Although these cells are circular because of the nature of radio signals, which radiate in all directions from a single source, they are usually represented on maps and in drawings as being hexagons, since that makes it easier to show where one cell ends and another begins; see Figure 1.1.)

Each cell has at its center a cell site, where the fixed radio receiver and transmitter are located. All the cell sites belonging to a particular system are connected together at a mobile telephone switching office (MTSO), which ties them into the conventional phone system. The transmitter at the cell site (and the one in your car) is low power (100 watts or less), and the effective useful radius of a cell is only a few miles. When you approach the working limit of one cell, your call is transferred, or handed off, to a cell site closer to you that can "hear" you better. (Figure 1.2).

FIGURE 1.1
CELLS ARE REPRESENTED BY HEXAGONS

Cellular service areas, or cells, are actually circular in shape because of the nature of radio waves.

However, they are usually represented by hexagons, since that makes it easier to show where one cell ends and another begins.

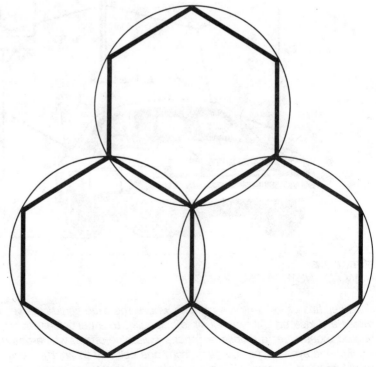

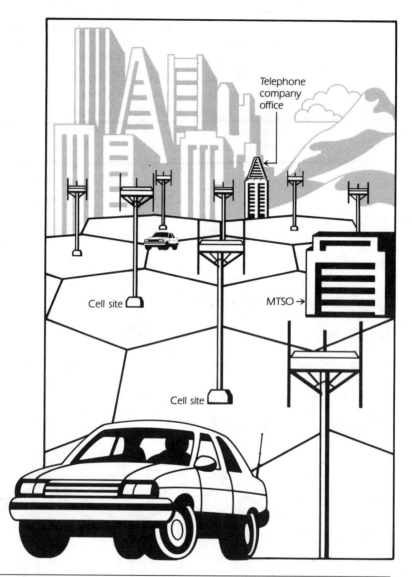

FIGURE 1.2
HOW CELLULAR PHONE SERVICE WORKS

Each cell has at its center a cell site where the fixed radio receiver and transmitter are located. All the cell sites belonging to a particular system are connected together at a mobile telephone switching office, which ties them to the local phone system. As you pass from one cell site to another, your call is transferred or "handed off" to the next cell without any noticeable interruption.

This limited-range cellular approach offers quite a few advantages over those used for earlier mobile systems. A number of small cells means that when you are within an area of cellular service you are always assured of strong signals—as you approach the limit of the range of one cell, you will enter another before the quality of service deteriorates. Furthermore, the limited range of cellular transmissions means that there will be no interference from other neighboring systems and that the same frequency can be used again just a few cells away. This is called *reuse* and makes it possible for cellular to provide greater capacity within a single metropolitan area. Finally, the cellular approach allows for easy expansion of service. When a particular area becomes saturated to capacity, a cell can be split to form several new cells. This can be done by adding directional antennas to a cell site so it can radiate several non-interfering signals in several directions at once, on frequencies different from the ones used by the original site. Each

Now you can stay in touch with office, clients, and family while driving in commuter traffic.

newly created cell can handle up to about sixty channels. Newer techniques include "microcells," which provide even smaller cells to reach hard-to-cover areas and still more reuse and capacity.

In the chapters that follow, we'll examine in greater detail how cellular systems—and, in particular, cellular phones—operate and present you with the information you'll need to take advantage of the many pluses of cellular communications.

2 Cellular Phone Equipment

There are three parts to a cellular phone system: the installation in your car (or your self-contained portable phone); the cell site, which receives and transmits radio signals from and to your phone; and the MTSO, or mobile telephone switching office, which links mobile phones to established conventional telephone services. The relationship among these is illustrated in Figure 2.1.

Of these three, the most important to you is the first, the cellular phone itself.

A CELLULAR PHONE INSTALLATION

A typical cellular phone installation consists of four basic components: the power source, the control head, the transceiver and logic unit, and the antenna. Most problems with installed car phones are a result of poor or improper installation.

Power Source

In most cellular phone installations the phone receives its power from an automobile battery. If it is part of a permanent installation, the phone may be permanently connected, or hard-wired, to the vehicle's electrical system. In a portable system or one that is intended to be removed periodically—for security reasons or for use away from the car—

FIGURE 2.1
PARTS OF A CELLULAR SYSTEM

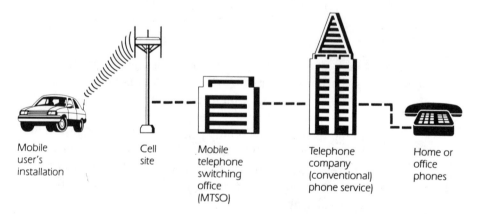

| Mobile user's installation | Cell site | Mobile telephone switching office (MTSO) | Telephone company (conventional) phone service | Home or office phones |

The three parts of a cellular phone system are (1) the installation in your car; (2) the cell site, which receives and transmits your calls to and from your car phone; and (3) the mobile telephone switching office, which then transmits your calls to conventional phone service.

a quick-disconnect plug or an adapter that lets the phone take its power from the car's cigarette lighter may be used.

Portable phones, of course, carry with them their own power sources, usually in the form of rechargeable battery packs. One-piece portables, which range in size from a brick to a pack of cigarettes, use *nickel-cadmium* or *nickel-metal* hydride batteries like the ones in such portable appliances as cordless electric razors and handheld vacuum cleaners.

Transportables—usually intended to do double duty both in a vehicle and away from it—may use a different sort of rechargeable battery called a *gelled-electrolyte battery* (gel cell). Gel cells are lead-acid batteries like the one in your

FIGURE 2.2
A TYPICAL INSTALLATION

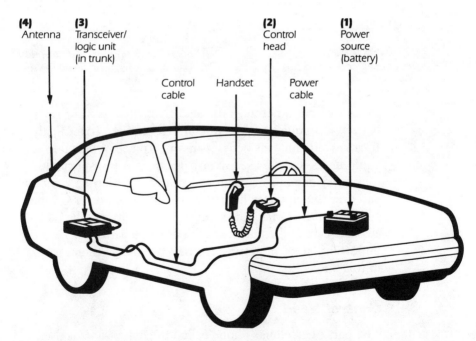

A typical cellular phone installation consists of four basic components: (1) power source, (2) control head, (3) transceiver/logic unit, and (4) antenna.

Portable phone with
self-contained battery.

Transportable phones are intended
for use in a vehicle or away from it.

car but with a difference: the acid electrolyte in your car
battery is a liquid, but the one in a gel cell is a sort of jelly.
It can't spill, and—as befits its use in a piece of portable
equipment—can be used in any position. A transportable
that is designed primarily for operation in multiple vehicles
and uses the cigarette lighter as its source of power is called
a *transferable*. It may not include its own battery. We'll discuss the care and feeding of each of these batteries, an
important factor in getting the best use out of them, later
in this book.

Control Head

The part of a cellular phone called the *control head* is used
to issue commands to the rest of the system, including the
parts physically close to you (such as the rest of your

The control head issues
commands to the rest of
the system. It may be built
entirely into the handset,
or may consist of two or
more pieces such as a
handset and a small
console.

phone) and those at the cell site. It is the part of the system
you most frequently come into contact with since it often
includes the telephone handset. The control head may be a
one-piece unit, built entirely into the handset, or may consist of two or more pieces such as a handset and a small console containing the phone's keypad and loudspeaker.

The control head is connected to the cellular phone's transceiver/logic unit by a multiwire cable that carries audio and control signals between the two.

Transceiver/Logic Unit

The heart of a cellular telephone is its *transceiver/logic unit.* You may think that the handset is the core of the system, but the real action is in the transceiver/logic unit. A *transceiver* is a combination radio transmitter and receiver (hence its name), and it is the means by which signals travel between your phone and the cell site.

In a permanent automobile installation, the transceiver/logic unit is mounted in the car trunk.

The logic unit is the part of the phone that contains the "smarts." It communicates with the equipment at a cell site to establish a connection, to determine what frequencies will be used for transmitting and receiving, and to coordinate its end of a hand off, when you leave one cell's area of coverage and enter another.

The intelligence built into a cellular phone's transceiver/logic unit is also used to control that phone's power output. If the equipment at a cell site senses that it is receiving a strong signal from a phone installation, it sends a signal telling the transceiver that it can cut back its power. Similarly, if the signal from a cellular phone begins to weaken as the phone is moved farther away from the cell site, it receives an instruction to increase its output. This ability to vary power output serves two purposes. First, keeping transmitter power to its usable minimum reduces the potential of a phone's signal getting into another cell site and creating interference. Second, reduced power output means a longer life for the batteries in handheld and transportable phones—more hours of operation per charge.

In a permanent installation the transceiver/logic unit is usually mounted in the trunk of a car and connected to the control head and car battery by cables. This location is chosen because it places this important piece of equipment in a presumably safe place and also keeps it from getting in your way. Such a location also places the transceiver close to the antenna, which is usually mounted toward the rear of a vehicle. This proximity allows the length of the cable that connects the transceiver to the antenna (the *feedline*) to be kept short. This is important when working with radio signals at the high frequencies used by cellular equipment, because the shorter the feedline, the better your reception will be. Also, the signal you are transmitting will be stronger. With the progressive reduction in its size, the transceiver may conveniently be mounted in a permanent or semi-permanent position in the passenger compartment, if desired.

The transceiver/logic unit part of portable cellular phones, or of the transportable phones intended to be operated away from a vehicle as well as in it, is either built into the all-in-one case of the portable or designed as a small unit that can readily be connected to (and disconnected from) a vehicle's antenna and battery. It frequently includes a means for semipermanent mounting near the driver's or user's position.

Antenna

Without an antenna, your cellular phone would be useless, since it would have no means to transmit or receive signals to or from a cell site. Several different types of antennas are used with cellular phones, and they will be discussed in detail later, in Appendix A. Because of the frequencies involved, it is possible to make cellular phone antennas quite small—sometimes less than a foot or so in length.

WHAT YOU SHOULD KNOW ABOUT CELLULAR FREQUENCIES

The frequencies used by cellular telephones range from 824 to 894 megahertz (MHz), with a gap between 849 and 869 MHz that's used by other communications services. (A map of the 800-MHz cellular phone spectrum is shown in Figure 2.3.) These frequencies were originally assigned to the top portion of the UHF-TV spectrum and were intended to be used by TV translator services for relaying commercial TV signals to small rural communities that had difficulty receiving regular broadcast services. The cellular phone frequencies are divided into two bands, and each band is subdivided into two sets of adjacent blocks, *A* or *B*.

Each area of cellular service is intended to be serviced by two companies—wireline service (a telephone company affiliate that usually already handles the existing hard-wired home and business service) and the other a *nonwireline* service (one that is usually already involved in other types of mobile radio communications or that operates a paging service). The differences between these two types of services are discussed in greater depth in Chapter 3, "The Business of Cellular Phones."

FIGURE 2.3
CELLULAR FREQUENCIES

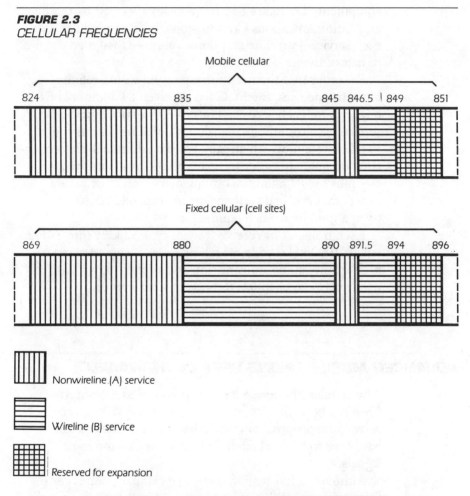

*Cellular frequencies range from 824 to 894 MHz, with a gap between 849 and 869 MHz for use by other communication services. The frequencies are divided into two bands, and each band is divided into two sets of adjacent blocks, **A** and **B**. Half of each band is assigned to a nonwireline (**A**) service, and the other to a wireline (**B**) service.*

One block of each band is assigned to a nonwireline, the *A* service, and the other to a wireline, the *B* service. (The *A-B* mechanism will be discussed further in Chapter 7.) The lower band (824-849 MHz) is for use by cellular phones, and the upper one (869-894 MHz), for cell site equipment. There are 832 frequencies allocated by the Federal Communications Commission (FCC) for cellular use; each service—wireline and nonwireline—is assigned the use of half of them.

Knowing that the need for expansion would require more frequencies, the FCC has reserved 83 additional frequencies for each service. Thus, the *A* service and the *B* service have room for expansion. Global represents the Pac Tel system, which already has the additional 83 frequencies "on line." In order to access all 333 standard frequencies plus the 83 additional frequencies, you must have a phone that is called "full-spectrum" capable. Global features a full line of such equipment.

Each transmit or receive frequency is 30 kHz (one MHz equals 1000 kHz) wide, which provides plenty of room for high-quality audio transmission and reception, with guard bands between channels to prevent interference.

ADVANCED MOBILE PHONE SERVICE STANDARD

The cellular phone service is also known as Advanced Mobile Phone Service, or *AMPS*. When AMPS was conceived, the proposal also included specifications for the hardware to be used, including connections and logic signals.

Although a few manufacturers of cellular phones conform to the AMPS standard, some do not. But whether the phone you use conforms to the AMPS control-head-to-transceiver-interface standard or not needn't really concern you—what you're interested in is how well the phone works, which is another matter altogether. However, the AMPS interface standard may be important to the people who install and service cellular phones; the type of equipment they use and the types of tests they can carry out depend on what internal signals a phone uses and on how easy it is to access and generate them for testing purposes. Let the installers worry about AMPS standards; it's their job.

NUMERIC ASSIGNMENT MODULE

One of the most important and most interesting parts of a cellular phone is a small integrated circuit, or chip, called a *NAM*, for numeric assignment module. (NAM rhymes with "Pam.")

The NAM chip is programmed, usually by your cellular phone dealer or installer, to contain the information that uniquely identifies your phone to a cell site when you place a call or when someone is trying to reach you through the cell site.

The process of NAM programming, sometimes called *burning*, requires special equipment, and once information has been entered into a NAM it usually cannot be changed (except with the same specialized equipment). Some NAM programmers are specialized "dedicated" devices, and some are intended to be used with a personal computer.

Included in the information programmed into a NAM is the serial number of the equipment you own or are leasing and the phone number assigned to it. The computers at cell sites and mobile telephone switching offices use this information to identify you when you call in—which, among other things, helps the phone company in preparing its bills—and to locate you when someone calls you.

The information contained in a NAM personalizes the equipment that contains it, which also makes it useful in identifying the owners of stolen cellular phones that have been recovered.

Quantum's pocket NAM programmer encodes the information that identifies your phone to the cell site.

A QUICK VISIT TO A CELL SITE

A cell site, which may cost a million dollars or more to construct and equip, is the link between your cellular phone and the rest of the phone system. It is where the messages bound for you leave the ground, as it were, and is the first stop for calls coming from your phone.

To get the best coverage, a cell site is sometimes located atop a tall building in metropolitan areas or on a high point or mountain in less built-up areas (Figure 2.4). This allows it to have the best radio "view" of the territory it is responsible for. The most distinguishing feature of a cell site is its array of antennas. Unlike your mobile phone installation, a cell site usually uses several antennas, each beaming and listening for signals in a different direction. These directional antennas ensure optimum results within a cell's area of coverage and may be individually adjusted for best results.

Each antenna actually consists of two antennas—one for transmitting and one for receiving. A single such unit is capable of handling any number of two-way conversations on different frequencies simultaneously.

Inside the cell site are the transmitters and receivers connected to the antennas, along with the equipment necessary to monitor the operation of the cell site and keep it in proper working order. Also located at the cell site are the electronics that connect the cell site to the mobile telephone switching office (usually by coaxial cable, microwave, or fiber-optic link).

FIGURE 2.4
CELL SITES NEED A GOOD RADIO "VIEW"

Cell sites are sometimes located atop a mountain or tall building to gain the best radio "view" of a territory.

A cell site can handle up to about sixty channels, based on using the same frequency no closer than seven cells away ($7 \times 60 = 420$, or approximately the 416 frequencies allowed one carrier). These same channels can be reused by nonadjacent cell sites within the same area, providing greater communications-handling capability while reducing a cause for interference between neighboring sites. For example, if you are using channels A and B at cell site one, the mobile phone users in the cell site next to you will use C and D. But, the people *seven* cell sites over will be able to "reuse" channels A and B, since they will be far enough away from you to do so without interference.

As cellular service continues to grow, more and more cell sites appear at shorter intervals, often down to less than one mile apart in densely-populated areas of cities. Where cell sites are closer to each other, there is less need for height. Rather, the antenna is kept low to reduce interference with other cells.

While height reduction, careful zoning, and careful design are used to prevent the antennas from becoming eyesores, they are becoming numerous. They are easy to spot because of their characteristic triangular array.

3

The Business of Cellular Phones

The business of cellular phones is a big one. Although cellular phone companies must make a large initial investment, the potential for an enormous return makes it certain that there will be no shortage of firms to service the cellular user.

WHERE CAN YOU GET CELLULAR SERVICE?

Cellular service was deployed beginning in 1983 in order of city population, with licenses for the largest cities being awarded first.

While some potential users have had to wait for service in smaller cities, in 1992 the last tertiary markets established commercial service. Now, cellular service is available in all populated areas in the United States (including Alaska and Hawaii), Canada, Mexico, and Puerto Rico. Only the most desolate wilderness areas do not have complete coverage.

HOW LICENSES ARE GRANTED: WIRELINE AND NONWIRELINE COMPANIES

The FCC, which controls and watches over the cellular communications industry, is responsible for issuing the licenses required to operate cellular systems.

To encourage competition, the FCC has decreed that provision be made for each area to be serviced by two cellular phone carriers, one a wireline and the other a nonwireline service. The *wireline* carrier is usually affiliated with the organization that already provides conventional telephone service—your local telephone company. Such companies are usually already involved in mobile communications in areas other than cellular and frequently bear a name similar to that used by their parent organizations.

The competitive edge required by the FCC in the cellular market is provided by another sort of company, a *nonwireline* service. This company is usually not directly involved with supplying standard phone service, but is likely to be involved in providing two-way mobile communications in its region of service.

Because several nonwireline companies have been purchased by wireline companies, the lines between the two are becoming blurred. In the case of a wireline carrier, there is usually little or no competition, and an application for a license can be evaluated and approved in a relatively short time. Wireline companies are encouraged by the FCC to enter into agreements among themselves regarding areas of coverage, responsibilities, and so forth. This simplifies their licensing procedure.

Who was to be an area's nonwireline carrier was another matter. The demand to provide service was so great that there were sometimes thousands of applications filed in just one month by nonwireline applicants. It was difficult for the smaller companies involved to reach agreements like those arrived at by the bigger wireline carriers. To avoid the delays in establishing cellular service that the resulting conflicts might have caused, the FCC chose to use a lottery system to award licenses to nonwireline services. The times these lotteries were announced and held was determined by a market's priority and by its need for service.

Most regions were usually first served by a wireline service and later by a nonwireline one. In some places, though, the nonwireline service was the first on the scene.

WHO SELLS CELLULAR PHONES

Cellular telephone equipment is available through many types of dealers—from stores specializing in selling telephones and electronic equipment to the local telephone company itself, through its authorized agents. Phones are available through the companies that provide nonwireline service and even from companies that specialize in phone installation. Some of these may offer discounts on certain lines and, in areas where the demand for service is high, may even have phone numbers preregistered and waiting.

In addition to selling you your phone and arranging for service to be started, most dealers either will install the equipment they sell or will refer you to someone who specializes in its installation.

The best way to find out about cellular facilities in your area is to ask the people who own and use the phones. They'll be proud to show off them to you and will be a rich source of information on phones, service, and the cellular situation in general.

Much more on the subject of selecting and purchasing a cellular phone system will be found in Chapter 4, "Getting Cellular Service."

4 | Getting Cellular Service

Cellular phone equipment must meet stringent standards and pass a number of stiff tests, so you are assured of getting a quality product no matter where you buy it. The real deciding factor on where to buy your cellular phone is the human one—the standards, know-how, and installation/service capabilities of the dealer.

FINDING THE BEST SOURCE OF CELLULAR PHONES AND SERVICE

Cellular phones are a hot item. In addition to Yellow Pages advertising, you can look in any newspaper or business magazine; ads for cellular equipment and services abound. Advertising for cellular service also appears on radio and television, particularly if you subscribe to a cable service that carries commercials. When purchasing a cellular phone, you should strongly consider dealing with an authorized service center of the product to assure factory-trained installation and service know-how.

EQUIPMENT COSTS

The price of cellular phones has fallen dramatically since they were introduced. The type of system that originally sold for several thousand dollars or more is now available for a price that ranges from as little as $595 to as much as $2,900, depending on the type of phone and the features you want. Because some dealers use their sales commissions to lower the price to the consumer (where it is not illegal), the price of the cellular phone can be almost zero, with $200–300 common.

Portable phones are generally more expensive than those intended for fixed mounting. Some fixed phones can lead dual lives, and transportability can be added for just a few hundred dollars. Others, particularly the more compact one-piece ones, sell for higher prices anywhere from $200–$1,000, with older models available at the lower prices.

As with anything else, the prices you see advertised may not include all of the necessities for the system such as antennas, cables, rechargers (for portables), and installation.

WHAT TO LOOK FOR IN A CELLULAR PHONE

Cellular telephones, because of their microprocessor-based design, include a number of features that make them more than just a way to talk to someone from your car or from the golf course. These hardware and performance features will be discussed in more detail in later chapters. For the moment, let's examine a checklist of things to bear in mind when shopping for a cellular phone.

First, make a complete list of all the expenses that will be involved in operating your phone. These include the cost of the phone itself, the accessories (including such basics as an antenna), the installation, and the subscription to and use of a cellular service. The next chapter presents some information about operating costs.

Make certain that repair facilities are readily available for your make of phone. This is particularly important should you need service; you won't be very pleased to find out that your phone has to be shipped halfway across the continent to be repaired or replaced.

If you live, or intend to use your phone, in a cellular service fringe area, note that some cellular phones have more-sensitive receiver sections than others (with the exception of handheld phones, the transmitter power is the same in all). Try out the phone you're interested in and compare it with others to verify that it will serve you well and dependably in your anticipated area of use. Most dealers will give you a demonstration if you're serious about buying. Take a drive through the area where you expect to be using your phone to make certain that you can maintain a reliable connection in "spotty" areas.

RENTING, LEASING, OR BUYING

Whether to rent, lease, or buy a cellular phone depends on your particular needs. The following sections will help you make the best choice for you.

Renting

In large cities it is possible to rent a cellular phone—usually the portable type that does not require installation—on a daily, weekly, or monthly basis. Prices vary. Renting can provide you with cellular phone facilities in an emergency or on a one-shot basis.

Do not judge the merits of cellular phone equipment on the basis of what you rent. Like any other rental item, the phone has probably been through a number of hands before yours and has almost certainly been subjected to a lot of abuse. (The battery packs of portable rental phones are notorious for their failures. This is not the fault of the batteries but rather of neglect in their maintenance and charging. See Chapter 8, "Portable and Transportable Phones," for advice on how to extend the life of the batteries in your own portable phone.) Renting is usually a short-term, desperation measure, and just as you would not want to own a rental car, you should also avoid owning a rental phone.

Leasing

Leasing a cellular phone, while more expensive than an outright purchase, can have a number of advantages. Not only are you assured of convenient repair facilities, but some leasing companies can provide you with a "loaner" or replacement phone should the one you are using go on the blink.

Leasing agreements usually contain a number of fringe benefits, the costs of which are built into the payments you make. These can include such things as

- Installation charge included in the initial fee
- A certain number of minutes of free usage, divided into peak and off-peak time
- Reduced rates after the minimum usage requirements have been met
- An extended (usually for the term of the lease) parts and labor warranty
- Free loss and theft insurance

Different types of leasing plans are available for business and personal use. The personal plans are generally less expensive but may not be as comprehensive as the ones for business phones, nor may they include all the benefits mentioned above. Investigate leasing plans carefully to compare the features of the different ones being offered in your area.

After the lease period has expired, you own the phone you have been using. Leases can, of course, be canceled. Bear in mind, though, that there is usually a penalty to pay, generally a forfeiture of your deposit ($200 or so). Also, some companies may charge an additional fee at the end of a leasing period when you decide to own the phone.

Because of the rapid decline in cellular phone prices, leasing is not as attractive as it used to be, and there is far less risk in outright purchase. Nevertheless, it can still be an attractive option, epsecially when the latest models and options may drive the sale price above $1,000.

Buying

Most people prefer to buy their cellular phones. The initial investment will probably still be smaller than that for a leased phone and, once it's paid for, the phone is *yours*, no strings attached.

Ownership carries with it certain responsibilities, however. If your phone breaks down—and even the most reliable equipment is susceptible to failure—it will have to be repaired. If you have a reliable dealer, this should not be difficult. Most cellular phone failures are simple to diagnose and set right; the problem-determining procedure is largely computerized, and repair usually consists of unplugging a defective internal module and replacing it with a good one.

Before you buy, find out if your dealer will service your equipment when the time comes and what the turnaround time will be. Make certain he is an authorized repair center for the brand you're buying—this will help guarantee that there will be replacement parts on his shelf. If you bought a bargain phone, there may be no local agency qualified to service it, and it may have to be shipped back to the factory or to a factory-authorized service center in another state.

Remember that if the phone you own is temporarily out of commission a rental phone can provide you with service. With call forwarding you can even have calls automatically transferred from your usual number to the one assigned to the rental phone.

Cellular phones come in a variety of styles.

Overall, outright purchase is probably your best bet. You will benefit from a lower price and from the other factors involved in ownership (availability of repairs, types of phone service plans, installation costs, and so forth).

TRYING IT OUT

Before you buy a cellular phone, you will probably want to try one out and make sure that the dealer has a reasonable return policy in the event that the phone does not work as promised. Dealers with whom you are seriously talking business will be happy to take you out and give you a demonstration, but to really know and appreciate cellular service and the convenience it affords, you have to live with a cellular phone for a while.

Renting can give you a taste of cellular service, but leasing a phone for a month or two will let you see just how valuable it can be to you. Should you decide to cancel the lease, you can write off the lost deposit to education; the

Try out the phone you're interested in to see that it will serve you well in your area and that it has all the features you want.

chances are that once you have experienced the convenience of cellular communication, you will not be able to do without it.

SPECIAL PROMOTIONS

By carefully shopping around you can make a cellular phone more affordable. This applies not only to finding the best prices on equipment but also to finding the best service.

Bargain phones almost certainly include nothing more than the cost of the phone. And although the extras (such as an antenna) may cost no more than if they were included in a package, their costs must still be taken into account when calculating your total investment. (Offers that claim to include a working phone number mean only that you will not have to wait for a number to be assigned to you; be assured that the cost of phone service is *not* included in the price of the phone.)

Another reason for bargain prices is that new models may have made the older ones "obsolete" and thus less expensive. But in a business as young as cellular telephones, it is difficult to imagine anything being truly outdated. The newer models may be a bit lighter or more compact or have a feature that was lacking on the previous ones, but the older models work just as well.

You can frequently get a deal on cellular service, too. Since the FCC has been thoughtful enough to allow for *two* cellular carriers in each service area, you may be able to take advantage of the competition between them. The incentives offered in carriers' promotions can include items such as discounts on regular and off-peak service, and a number of minutes of free usage. Keep your eyes and ears open and you may find someone making you an offer you can't refuse.

INSTALLATION

If you lease a cellular phone or purchase it from a cellular carrier, installation in your car will probably be included in the price. If you purchase the phone elsewhere, you may have the option of having it installed or doing the job yourself. Although installation is not an especially difficult task, you'll be most assured of getting a professional job that will serve you well if you entrust the job to a reliable installer.

The typical installation will cost in the vicinity of a hundred dollars and go up from there. Having someone else do the job for you has several advantages besides not getting your hands dirty. You get the benefit of the installer's investment in test equipment and of his expertise in using it. A good installer will not only put your phone in place but will also check it out thoroughly to make sure that everything works properly.

If your dealer does not do his own installation, he almost always has an arrangement with one or more qualified establishments who do the work for him on a regular basis. If you elect to find your own installer, the Yellow Pages can again assist you. Most of the firms you'll find there are reliable.

5 The Bill, Please

The costs of a cellular phone, of other necessities such as an antenna, and of installation are one-time expenses. Once you've paid for these items and services, you're done with them. Similarly, if you lease or rent a phone, you know what the charges are going to be and can easily budget for them. These fixed expenses can be more or less disregarded in day-to-day use of your phone.

Fixed costs, though, account for only a part of what it costs to operate a cellular phone. In this chapter we'll look at the other expenses—what it costs you to *use* your phone.

ONE-TIME EXPENSES

Like the price of purchasing equipment, some operating expenses are paid once and never again (unless, for some reason, you incur them again, as by having your phone number taken out of service and then requesting that service be restored). These one-time expenses include the following.

Service Activation Charge

This fee, typically about fifty dollars, is the phone company's charge for making the electronic arrangements for service to your number and for processing the paperwork associated with it. If you suspend service and then request that it be resumed, you may have to pay this fee again.

Change Charge

If you ask that your service be switched from one phone to another using the same number (maybe you just bought a new model), you will have to pay for the expenses incurred in making the switchover. Change charges are usually around fifteen dollars.

Deposit (Leased Equipment)

This is a charge to ensure that you will uphold your end of the bargain. If you don't or if you cancel your lease, you may have to forfeit this amount—usually several hundred dollars—which would otherwise be returned to you at the end of the lease period.

RECURRING EXPENSES

Just as with your home phone, the use of a cellular phone brings with it certain fixed charges, usually the same amount in every billing period. The following are among the charges that repeat from month to month.

Access Charge

This is the fee for basic service and, as explained below, may or may not include a number of free minutes of calling time.

Minimum Usage Fee

This is usually a part of the access charge. You may be billed for a certain number of minutes of use, regardless of whether or not you meet this minimum.

Special Services

If you have opted to add to the basic cellular services some of the options offered by your cellular carrier, you'll be billed for them on a regular basis. Interestingly, the cost of these additional conveniences (about three dollars apiece) is frequently less than that of the equivalent for a conventional phone. These extras can usually be added to your regular service at any time.

The following are among the additional services available.

Call Forwarding This service permits you to have calls transferred automatically from your cellular phone number to your home or office phone. If you have call forwarding you can switch it on and off from your cellular phone simply by pressing a couple of keys. A similar feature, called *no-answer transfer,* allows calls to your cellular number to be transferred to your home or office phone if you do not answer after a number of rings.

Call Waiting This service gives you, in effect, a second line to your cellular phone. If a call is placed to your cellular number while you are already using your phone, you will be alerted by a beep from your phone's speaker. You can put the first party on hold to speak to the second, or vice versa. It's just like having a second line on your phone, hold button and all, the difference being that there is only one phone number involved, not two. Like call forwarding, call waiting can be switched on and off with a couple of key presses.

Local Calls Only This feature prevents your phone from being used for long-distance calls—calls can be placed only within your calling area. (This can be a useful feature if people other than yourself use your phone.)

Incoming or Outgoing Call Restriction This permits you to make your cellular phone *receive only* (people can call you, but you can't call out) or *send only* (you can call out, but no one can call you).

VARIABLE CHARGES

As you might expect, the more you use your cellular phone, the greater your monthly bill will be. The way you are charged for service will vary from one part of the country to another and possibly from one service to another. The charges and billing methods described here are intended to be typical of what you can expect; they do not represent the rates or policies of any specific carrier.

Most companies offering cellular service have several pricing plans that you can select from. Some of these include in the cost of basic service a number of minutes of calling time; other plans separate the two. Some plans carry different rates for usage at the same times—one plan may be designed for people who make the greatest use of the phone during business hours, while another may be aimed at nonbusiness users who place the majority of their calls in the evening or on weekends.

Base rates, sometimes called *access charges,* start at about fifteen to thirty dollars per month and may go as high as fifty dollars, with no free time included (see Figure 5.1). Generally, the rate per minute of connect time goes down as the access charge goes up—with things tending to balance out in the end. When you're shopping for a service, explain how you intend to use your phone and the sales representative will suggest a plan appropriate for you.

As an example, a service may have a plan with a forty-five dollar base rate, with calls during peak time costing forty-five cents per minute and off-peak calls costing twenty-seven cents per minute. (Peak time in many areas extends

FIGURE 5.1
A TYPICAL CELLULAR PHONE BILL

WESTERN MOBILE SERVICES
SAN FRANCISCO
BILL SUMMARY

BILLING DATE 10/10   BILLING ACCOUNT #14002487

PREVIOUS BALANCE	46.63	
PAYMENTS	46.63	THANK YOU FOR YOUR PAYMENT
BALANCE		0.00
MONTHLY ACCESS		25.00
OPTIONAL FEATURES		0.00
AIRTIME USAGE		65.00
LONG DISTANCE		0.00
PACIFIC BELL		00
ROAMER TOLL		3.77
OTHER CHARGES AND CREDITS		0.00
LATE PAYMENT		0.00
FEDERAL EXCISE TAX		2.80
STATE 911 TAX		0.83
LOCAL TAXES		0.00
STATE REGULATORY FEE		0.27
BILLING SURCHARGE		0.00
TOTAL AMOUNT DUE		**97.67**

from 7:00 a.m. to 7:00 p.m., Monday through Friday. The other twelve hours of the day, as well as all day Saturday and Sunday and some holidays, are considered to be off peak.) Another plan offers service with a twenty-five dollar base rate, with calls during peak time costing ninty cents per minute and off-peak calls costing twenty cents per minute. Depending on your pattern of use, one plan or the other will be more to your benefit.

Some budget and most leasing plans include a number of free minutes, but, as a rule, the lowest rates require you to pay extra for all your calls.

There are usually three to five types of rate plans, each of which may be attractive based on usage or longevity. The basic plan is for average users who conduct some business on the phone. Other plans, intended for "heavy" users, discount both the usage rate and the access charge, based on the user's commitment to a minimum usage amount and/or a minimum period, usually one year.

Other business-oriented rate plans may provide a discount for a single company with multiple users billed to one address. Very light users and non-business users may benefit from a plan that has a very low access charge, high per-minute charges for off-peak use, when they may do most of their calling.

Periodically, a carrier will have a special promotion on the service, in the same way that a cellular telephone dealer may have a promotion on cellular phones. The carrier may waive the service activation charge, provide an allowance of minutes of free usage, or provide premium services like voice messaging, special services, or paging free for a trial period.

HOW CALLS ARE BILLED

One thing that differentiates cellular charges from those of home and office phones is that no matter whether you make a call or receive one, *you* pay for it.

When you place a call, the meter starts running when you press the SEND button on your phone to initiate contact with the cell site. Thus you are billed from the time you start to place your call, not just the time you are connected with the other party (see Figure 5.2). In the world of cellular phones, *connect time* refers to the time you are connected (by radio) with the cell site and the mobile telephone switching office (MTSO), not with another phone. The calling period ends when you either hang up your phone or press your phone's END button, both of which terminate radio contact with the cell site.

When someone calls you at your cellular number, you pay for that too. The billing period starts when you answer and ends when you hang up or press the END button.

Cellular calls are often billed by the minute or part of a minute. A call lasting two minutes and one second will cost the same as one lasting two minutes and fifty-nine seconds. In some areas charges are billed by the nearest tenth of a minute. No matter whether you place a call or receive one, it is you—at the cellular phone end of the conversation—who pays the bill.

LONG DISTANCE AND ROAMING

From your cellular phone you can call—and receive calls from—anywhere in the country or in the world. The cost for a long-distance call is not much different from what you would be billed for from your home or office phone. In fact, the price is exactly the same except that the normal cellular connect charges are added to it. Therefore, a call that would cost you, say, seventy-nine cents a minute if you made it from your office, might cost you one dollar and nineteen cents a minute (seventy-nine cents plus forty cents) when placed from your cellular phone.

In some areas, using your cellular phone will save you money because a long-distance call from your office or home phone will be a local call using the cellular service.

FIGURE 5.2
HOW A LONG-DISTANCE CALL IS BILLED

(1) Calling long distance

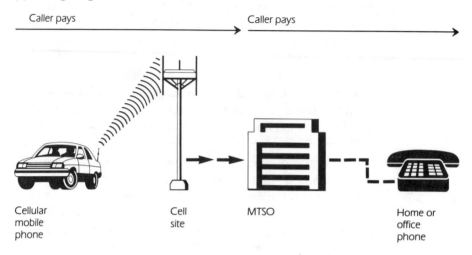

(2) Receiving long distance

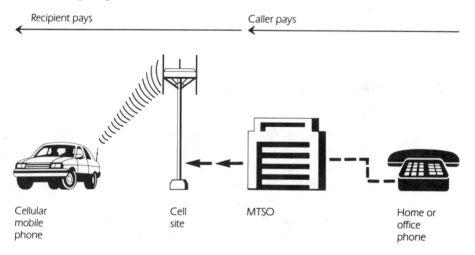

*(1) When you place a call, you pay the long-distance rate between the MTSO and the point you are calling. (2) When someone calls you, **they** pay the rate between their phone and the MTSO, and **you** pay for the portion of the call between the MTSO and your cellular phone.*

For example, calling from Orange County, California to Los Angeles would be a long-distance call using your regular office or home phone. But, using the cellular service, these two areas are part of the same local calling area, so you will be billed for a local call.

When you make a long-distance call, you pay the going long-distance rate between the MTSO you are using and the point you are calling, plus the standard per minute cost of cellular service. If people call *you* long distance, they pay whatever long-distance charges normally apply between their phone and your MTSO (generally the same as to your home or office number), while you are billed for the portion of the call between the MTSO and your cellular phone.

Roaming refers to the practice of using your cellular phone when you're outside your local calling area (and therefore using another cellular service provider's facilities). This is such an important part of cellular life that it has its own chapter—Chapter 7—devoted to it. There is much confusion about how roaming works and how it is charged for that will be cleared up in that chapter. Suffice it to say here that there are usually special charges applicable to roamers.

CHOOSING A SERVICE PROVIDER

Since there are two carriers in each area, you may want to consider the differences in service providers. In addition to the two carriers, some cities have *resellers*, which buy service at wholesale from one or both of the two carriers and sell it at retail to end users, increasing your choices. Service providers therefore include resellers and carriers.

Note that the dealer who sells you a cellular phone may have an exclusive arrangement with one service provider and may be able to initiate service for you with that one only. You may like the service, installation, and price you get from the dealer on the phone, but then want to investigate service providers separately.

In addition to pricing differences (in both the price level and the structure of the rate plans) and periodic promotions, service providers may be different in many other characteristics. One may have less service coverage than the other. One may have a larger area in which usage charges include a toll.

Because systems can become congested with too many users, a service provider may have inferior service quality, characterized by dropped calls, several attempts required to put a call through, interference, static, and so on. And different providers may have a different mix of special and premium services available.

Finally, customer service, billing convenience and simplicity, and other considerations can make the difference. The best way to evaluate service providers, if you don't have special needs in these areas, is to talk to current users.

6 Hello, Ma?
It's Me!

Using a cellular telephone is not much different from using a more conventional one. In fact, it can be even easier because these phones contain their own small computers. And the high quality of their audio makes them a pleasure to use.

At first glance, the keypad of a cellular phone may appear intimidating. Once you know what everything's for, though, and how it works, there's nothing to it.

BUTTONS, BUTTONS

In addition to the usual twelve buttons found on a home or office touch-tone phone, the handset of a cellular phone contains several buttons that are intended specifically for cellular operation. Because convenience and safety are two prime factors in the design of a cellular phone, as many of its controls as possible—sometimes all of them—are grouped together on the handset. Figure 6.1 illustrates a typical handset.

The SEND key (1) puts the phone into action when you are ready to place a call. Until it is pressed, nothing will go over the air or into the phone system. This allows you to dial a number while parked or at a stop light and send at your leisure, without tying up your equipment or taking your eyes off the road. (You can still receive calls while you're dialing in this fashion.)

When the number to be called is available in the phone's working memory (discussed below), you can press the SEND button and you'll be on the air. The IN USE indicator will light to show you that a connection with the cell site has been established. You can fiddle with the keypad all you like at any time, but until you press the SEND key, everything you do will affect only you and your phone and not go out into the system.

The CLEAR key (2) helps you fix your mistakes. It is all too easy—especially if you disregard the advice given in this book and try to dial a number while you're in motion—to hit the wrong key. The CLEAR key allows you to erase the last digit you entered. If you want to erase an entire sequence of digits from the phone's immediate working memory—perhaps you got the first digit of the area code wrong and didn't realize it until you had entered the entire phone number and saw it on the phone's liquid crystal display (LCD) screen—with most phones you press the SEL key (more about that one shortly) and *then* CLEAR.

The purpose of the END key (3) is found in its name. It's used to hang up the phone. Sometimes you don't want to have to replace the handset in its cradle to terminate a

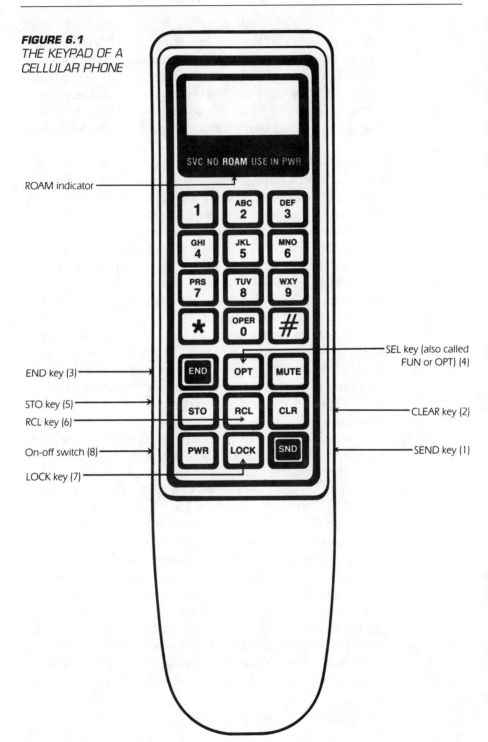

FIGURE 6.1
THE KEYPAD OF A CELLULAR PHONE

ROAM indicator

SEL key (also called FUN or OPT) (4)

END key (3)

STO key (5)

RCL key (6)

CLEAR key (2)

On-off switch (8)

SEND key (1)

LOCK key (7)

call, or you want to hang up in a hurry. The END key lets you do that just by pressing it. You can use it to end one call and can then place another without having to physically hang up the phone. You can also return the handset to its cradle to terminate a call, but you might not want to do that if you were in heavy traffic, hence the handy END button.

Several of the keys on the handset serve more than one purpose. This is done primarily to conserve space—if each of the functions available from a cellular phone had to be called up by a separate key, there wouldn't be room for them all on a handset of any reasonable size.

These extra functions may vary from phone to phone, but a number are found on just about every phone. To invoke these functions, you have to press a sequence of keys—one key after the other. Most functions require the pressing of only two or three keys.

The first key that must be pressed in all such cases is the SEL key (4)—it permits you to SELect the function you want. The next key (or keys) that you press is usually one of the numeric keys—those used in dialing a number. When they follow the SEL key, however, they take on a new meaning.

Some of the keys that are used for special functions bear secondary labels (Figure 6.1) that define their purpose; more rarely used ones require their functions to be memorized. The labeled ones may include STO, RCL, LOCK, and HORN (these are numbered 5-8 in Figure 6.1).

The STO key (5) allows you to STOre numbers (in some cases with as many as thirty-two digits!) in the phone's permanent memory for later recall. (On some phones this key is called the FUNction or OPTion key.) When you press the SEL-STO key sequence, you will be able to enter a phone number from the keypad. After checking the screen to make sure it's correct (and using the CLR key if you made a mistake), you press the STO key (by itself) again. The phone will then allow you to enter a one- or two-digit number that will recall and dial the longer one.

The RCL key (6) permits you to ReCalL a phone number from the phone's permanent memory by using the one- or two-digit number you assigned with the STO function. By pressing RCL and then the reference number, the entire phone number it represents will be brought into the phone's working memory and displayed on its LCD screen. Then by pressing the SEND key, you can initiate a call to that number.

The LOCK key (7) is part of a phone's security system. When you press it, the word LOCK will appear on the phone's display and it will be impossible to dial out, although the phone will still be able to receive calls. This feature is intended to protect you should you lend the vehicle containing the phone to someone or should the phone or the vehicle be stolen. In the latter case, you may no longer have the phone, but whoever took it will not be able to make calls on it.

The information to unlock the phone is usually programmed into its NAM, which also contains such pertinent information as the phone's serial number and the phone number registered to it. (This information is automatically transmitted to the cell site when you use your phone to permit the proper billing and, of course, to allow your system to recognize when it has a call for you.)

To unlock the phone, you must enter a sequence of numbers—perhaps four or five of them—known only to you. It may be your birthdate, part of your home phone number, or any other figure that is easy to remember. Never write down your unlock code where it can be easily found and associated with your phone—it might be discovered by someone who had no business having it and who might be tempted to use your phone without your knowledge or permission.

The use of the horn button is described more fully in the chapter on options and accessories. In brief, it allows you to use your vehicle's horn to alert you to incoming calls should you be out of the car but within hearing range.

PLACING A CALL

When you pick up the handset of a cellular phone, you don't hear a dial tone. That's because, until you press the SEND key, you aren't connected with the MTSO (where the signal originates) and also because, in cellular telephony, a dial tone isn't necessary. The purpose of a dial tone is to tell you that you are connected with the phone system's switching equipment; a cellular system has a different way of doing that.

You dial a number just as you would from a standard phone, by pushing the buttons corresponding to the digits. As you enter the digits, they will appear on the display above the keypad (Figure 6.1), each digit being bumped along from right to left as new ones are keyed in. (You can even dial a number the night before if you feel like it.) These digits will stay in the phone's working memory until you clear them out or until they are replaced by another number.

If the number you want is already stored in the phone's memory, simply press the phone's RCL button, followed by the appropriate code. (If you've forgotten the code—usually a number from 01 to 16 in the case of phones that can store sixteen numbers—there is usually a way for you to scan through the codes and numbers to see which goes with which.)

With the number now in the phone's working memory (indicated by its being shown on the display), you are ready to establish the connection by pressing the SEND button. This activates the phone's transmitter and sends the dialing information to the nearest (or strongest) cell site, from which it will be sent on to the MTSO. If you are successful in reaching the cell site—as you usually will be—the phone's IN USE indicator lights up. If, for some reason, your phone is unsuccessful in reaching the cell site, the NO SERVICE indicator will come on.

There are several reasons why you may get a NO SERVICE indication. The most likely is that you are not within range of a cell site. This may be the case even if the NO

SERVICE indicator is not lit when you check it. The reasons for this happening and the ways in which you can improve your chances of getting through under difficult conditions are detailed in Chapter 10, "Dealing with Operational Difficulties." You'll find that the conditions under which you can receive a call or maintain a connection in a "spotty" area are different from those required to place a call. After a little practice, you will learn to tell how long you should expect to wait for a connection to be established and what to do to improve your chances under less-than-ideal conditions.

If you are unsuccessful in establishing a connection the first time, try again by pressing the SEND button—the number is still in the phone's working memory even if it has disappeared from the display (it will reappear when you press the SEND button).

Unlike conventional phone systems, as your call progresses toward the point where you are connected with the number you've dialed, a cellular system lets you hear no tones, clicks, or beeps. What you will hear from your phone's speaker or earpiece is a silent hum, which is the carrier signal from the cell site. (It is like the sound you hear when a broadcast radio station is on the air but is not transmitting any speech or music.)

The first sound you'll hear from your phone will be one indicating that the remote phone is ringing . . . or that it is busy. In the normal course of events the next thing you'll hear will that of the remote handset being picked up and a welcome "Hello."

There are two ways to terminate a cellular phone conversation. One is to physically hang up your handset, as you would with a conventional phone. The other is to press the END button on the handset. This allows you to finish one call and immediately place another.

Remember that if *you* don't terminate the call, even though the party on the other end has hung up—whether you placed the call or just answered it—you're still connected to the cell site and paying for that connect time for the next twenty to thirty seconds. After that the MTSO will (in most markets) automatically disconnect your call.

RECEIVING A CALL

Receiving a call on a cellular system is even easier than placing one—assuming that your phone is turned on and that you're within range of a cell site.

When someone dials the number of your cellular phone and is connected, your phone will ring (and the caller will hear a ring signal at his end). There is no bell in a cellular phone—the ringing sound is generated electronically and reproduced by the phone's speaker. All you have to do pick up the handset and answer or, if you have a hands-free

With a cellular phone, you're never far from important calls.

phone (see Chapter 9, "Options and Accessories"), press the SEND button. Don't forget that it's up to you to hang up!

SPECIAL CELLULAR FEATURES

Companies that provide cellular phone service offer their customers a number of optional features that can be added to the basic service. These features can tailor the service you pay for to your particular requirements. Some of these features were mentioned in the previous chapter; a more comprehensive list follows. Generally, these features are available at a monthly cost of about three dollars apiece. Some high-end subscriber plans may include a choice of one or two of these features at no additional charge.

Local Number Plan Area Calling Only

A local number plan area, or NPA, is what your cellular phone company calls your local (although it may include area codes other than your own) area of service. In the Los Angeles metropolitan area, for example, the NPA includes area codes 714, 213, 818, and parts of 619 and 805.

If your phone is used by a number of people—or if you are responsible for a fleet of phones used by your company's sales force—this feature can save you money by making it impossible for a phone to be used for unnecessary long-distance toll calls. Restricting the phone's usage to your local area in no way diminishes its effectiveness as a sales or business tool; it simply concentrates the phone's power in the required area. If you anticipate that all the calls from your cellular phone(s) will be local ones (this type of call restriction does not, however, block out incoming long-distance calls), the slight monthly charge can be a small price to pay to guarantee that the phone won't be misused.

Incoming-Call Restriction

This feature makes your phone an outgoing-calls-only system; you can only place calls, you cannot receive them. It is useful if you do not want to be disturbed.

Outgoing-Call Restriction

This feature is the opposite of the aforementioned one. The phone can receive calls, but none can be placed from it. Under some circumstances this may be a desirable feature since, if no calls can be placed from the phone, there will be no charges for them. (You'll still pay, of course, for the connect time for incoming calls.)

"Oh nooooo, Dad put the phone on outgoing call restriction, and we really must call Beth, then Pam, then Zack in New York, Margie in California, Nick in Chicago, Naomi in Boston, ..."

Speed Calling

This feature allows you to store a repertory of eight or so frequently called numbers at the MTSO and call them simply by dialing one or two digits. It is of dubious value since most cellular phones already include a similar feature.

Call Forwarding

If you're expecting an important call at your cellular number but can't stay by the phone, call forwarding allows you, by pressing a few keys, to have calls placed to that number automatically transferred to any other number (including numbers with different area codes). This feature can be turned off and on at will simply by entering a two- or three-digit code on the phone's keypad. This is a one-way feature. That is, it will only transfer calls intended for your cellular number to another. It will not transfer calls placed to a home or office number to your cellular phone. You have to subscribe to call forwarding separately for that number through your regular phone company.

Be aware, however, that even though a cellular phone is not directly involved in this type of call (since the call never reaches it) the usual per minute airtime charges may apply, as well as any others such as long-distance tolls.

Call Waiting

This feature gives you the equivalent of a multiline phone at your cellular number, allowing you to be reached even when you are already using your phone.

If people try to call your cellular number while it is in use, instead of a busy signal they will hear a normal ringing sound (this is eventually replaced by a busy signal if you ignore them long enough). Your phone beeps to alert you to a second caller. You can end the first conversation to begin the second, or you can alternate between the two.

Note that you will be billed for *both* calls. The billing period for the second call begins when you are alerted to it—that is, when your phone beeps.

Three-Party Conferencing

This allows a third party, from a third phone number, to be added to a conversation. You are billed for connect time to both other numbers.

No-Answer Transfer

This is similar to the call-forwarding feature discussed above. It differs from it in that the call intended for the cellular phone is not automatically transferred to another number right away. Instead, the cellular unit is allowed to ring for a predetermined period of time. If, after that interval, it is not answered, then the call is transferred to the other number (a call will also be transferred if the cellular phone is turned off or if it is out of range of a cell-site's transmitter).

This feature cannot be used at the same time as call forwarding and vice versa (although you can have both of them available to you). As with call forwarding, you are billed for the airtime used.

It can sometimes be confusing trying to understand when *you* pay for a call and when the other party does. A simple rule of thumb is to assume that you pay for anything that travels between your phone and a cell site, as well as for any toll charges that you incur in placing a call (say, to another area code). This also applies to calls that use call forwarding or no-answer transfer, even though your cellular phone is not directly involved in the call.

If, however, people call you long distance, they pay for the overland portion of the call—you're responsible only for the portion carried between your phone and the cell site you use.

7 Roaming

Using your cellular phone when you're away from your home cellular service area, traveling in another, is called *roaming*. Virtually all cellular carriers have made arrangements under which you can simply arrive in a service area other than your own and automatically start to use that carrier's service.

There is a certain mystery surrounding roaming, and a lot of questions have arisen over what roaming is and how it works. This chapter is intended to answer those questions.

HOW TO TELL WHEN YOU'RE ROAMING

The boundaries that mark the end of one phone company's domain and the beginning of another's are invisible. Unless you've traveled the route before, you can't tell when you've left your home area and entered another. But your phone can.

One of the indicators on a cellular phone is labeled ROAM (see Figure 6.1). It serves several purposes. When you leave your home region and enter another, the ROAM indicator will light and stay lit as long as you are within range of a cell site. If you are on your way from one city to another, the ROAM indicator will go out and be replaced by the NO SERVICE light as you pass through an area in which cellular service is not yet available.

The ROAM indicator serves another purpose. All cellular phones are equipped to switch automatically between two services, wireline and nonwireline. The service you normally use is programmed in the phone's NAM to give one priority over the other. This doesn't matter if there is only one cellular carrier in your region, but it's important if both a wireline and a nonwireline service are active. On many phones you can program which service is to have priority over the other and even lock out one or the other completely. (The mechanism that does this is called an *A-B* switch. *A* services are the nonwireline services, *B* services, the wireline ones. All cellular phones have *A-B* switches, but some are more versatile than others.)

It's advisable to switch your phone manually between *A* and *B* services as you travel, setting it to the one you expect to use. That way you won't inadvertently try to use a carrier that doesn't know you exist and that can't provide you with service.

If you leave your *A-B* switch in automatic or priority mode as you travel, your phone's ROAM indicator, instead

of lighting steadily, may begin to flash slowly. This tells you that the service your phone is "listening to" is a service other than the one it is programmed for—perhaps a wireline service when you normally use a nonwireline service. Usually, your phone will automatically make the switchover, and, except for the flashing light, there will be no difference apparent to you.

If you often ROAM to the same areas, there are phones available that hold multiple NAMs. This way you will be "local" even when you are roaming away from home, and as a result, will not be billed as a roamer.

The flashing indicator can be startling the first time you notice it. Don't be alarmed, there's nothing wrong with your phone. It's just keeping you informed of what's going on.

ROAMING: OUTGOING CALLS

If it weren't for the ROAM indicator and the changing scenery, you might never know that you were away from home, calling on a service other than the one you normally use.

It is important to be aware of the area code of the system you're operating from. If you want to place a call to anywhere other than your home service area—even if it's within the area you're visiting or passing through—you must use an area code! The only exception is if you call home. Then, no matter where you may be calling from, you do not need one.

Some cellular services have adopted a policy of requiring an area code for all calls, even local ones. This eliminates any possibility of confusion. If you omit this prefix where it's required, you'll get a recording telling you to dial again.

You may be accustomed to dialing the numeral 1 before an area code when calling long distance from your home or office phone. Some cellular systems do not require this—all you need is the three-digit area code, followed by the additional seven digits specifying the exchange and number. If you forget and use a leading 1, you may get a recording telling you to try again without it.

ROAMING: INCOMING CALLS

If you're within range of a cellular system, no matter where you started out from or where you may be, you can receive calls at your cellular number. It's a bit more complicated, however, than placing them.

The problem is that callers have to know where you are. If they call you at your cellular number and you are roaming in another cellular system, they will get a recording indicating that the local system can't find you—the same message they would get if you were home with your phone turned off. To get through to you when you are using the services of another system, a caller has to know what system you are using and how to access the facilities of that system.

The first part is easy—you leave an itinerary with the access numbers for the cellular systems in the areas you'll be passing through. A cellular system's *access number* is a phone number that a caller must use to call into that system from outside its area code. A typical access number might be (area code) *555-ROAM*. (A list of roamer access numbers for most major cities is shown in Appendix B. Many, but not all, systems use the numbers formed by the letters R-O-A-M.)

To call a roamer, first dial the access number of the system you expect him to be using. There will be a short pause and then you will hear a signal (usually a tone) that indicates you can now dial your roamer's number, home area code and all. The area code—the roamer's local one, not that of the region he's visiting—is necessary because the rest of his phone number may duplicate one that's already in use where he is. With that—which is a lot simpler to do than it is to read about—you should hear your roamer's phone start to ring.

This is the way things work for most areas. In some parts of the country, however, you must write or call ahead to make arrangements to use your phone during a visit. Before you go anywhere you should check with the phone systems you intend to use to determine their requirements. Your own cellular service—ask for the Roaming Coordinator or Roaming Supervisor—can frequently assist you in arranging to use another facility, including advising you about their requirements.

Most cellular companies have agreements that make pre-arrangements for roaming unnecessary. Others—and you can find out which ones through your Roaming Coordinator—require prior notification before you can use their systems to make or receive calls.

Automatic roaming is being replaced by "follow me" roaming. When a call is made to your cellular phone in your home area and your phone is off or out of the area, your carrier will check most carriers (either A or B block) for the presence of your phone and connect the call if it finds you. Remember that you must pay additional long-distance charges on top of roaming charges for such a call. Eventually, roaming will be entirely automatic, and all incoming calls will reach you.

When you have to make arrangements for roaming, you will be asked for certain information. The most important are your phone's Electronic Service Number (ESN), also known as its registration or serial number and, of course, the phone number assigned to your cellular unit. You will also be asked for the length of your stay and for credit card information. (Some service carriers will not have a roaming agreement with your local carrier whereby roaming charges will appear on your regular bill; in this case the charges must be billed to a major credit card.)

Don't be too concerned, by the way, if not all your roaming charges show up on one bill. The complexity of coordinating the billing of many separate companies sometimes creates delays, and airtime charges and long-distance charges, even for the same call, may not be billed at the same time.

With the advent of clearing houses, which settle the roamer billing for cellular carriers, and new measures to combat fraud, most roaming is virtually automatic, and manual arrangements to roam will rarely be necessary.

Roamer Fraud

Many different schemes have been devised to attempt to obtain free cellular service. In some cases, cellular telephones have been programmed with fictional telephone numbers from foreign cities and with the number changing often, a practice called *tumbling*. A more sophisticated scheme, called *cloning*, involves the interception of the ESN and telephone number combination and "cloning" the legitimate combination to another phone.

But as these schemes have been uncovered, new methods of securing the cellular system and checking the offered combinations against the home data base in real time have been developed. In the past, some carriers have temporarily disabled automatic roaming whenever fraud became significant, inconveniencing honest customers. Today, fraud has been reduced to near zero, and the user may assume that automatic roaming will be available everywhere at all times.

BILLING

With all this calling back and forth going on, you might be confused about who pays for what in a roaming situation. It's not difficult to sort out if you look at things one piece at a time.

A Roamer Calls Within a Roaming Area

This, of course, is the most straightforward case, where a roamer makes a call to a number located within the area he is visiting, which he can usually do just by dialing the number (including the area code).

Each cellular system has its own rates for roaming callers (the charges will appear on your regular statement or, as described above, on your credit card statement). Table 7.1 shows the rates charged in several cities. In most cases there are different rates for peak and off-peak service. The cost per minute to roamers is usually a bit higher than it is to local users largely because of the extra costs involved in billing. These per minute charges are applicable whether you place or receive a call as a roamer.

TABLE 7.1   ROAMING RATES FOR TEN MAJOR CITIES

City	Switch	Activation fee	Per day +	Rates Per message unit Peak	Off-peak
Dallas/Fort Worth	B			.37	.24
Los Angeles	B			.70	.27
Chicago	B		.75	.45	.45
	A			.35	.35
San Francisco	B		2.00*	.50	.50
Cleveland	A		2.50	.45	.30
	B		2.00	.50	.50
Denver	B			.60	.60
Detroit	A			.35	.35
	B		.75	.45	.45
New York City	A				
	B	10.00	3.00	.55	.55
New Orleans	A		2.00	.40	.40
	B		2.00**	.60	.60
Boston	A			.55	.55
	B	5.00	3.00	.50	.50

*No daily charge for GTE customers.

** No daily charge for Bell South Mobility customers.

A few systems charge for incomplete calls (such as those where a busy signal is received). There is usually a fixed fee for this.

A Roamer Calls Home

When a roamer places a call to his home calling area, he is billed the normal per-minute (roamer's) usage charge of the area he's in as well as the long-distance charges between the MTSO he is using and the number he is calling.

A Roamer Gets a Call from Home

When someone at home uses an access number to reach a roamer, the roamer is billed (per minute, as usual) at the standard roaming rate. The long-distance charges are billed to the calling party.

A Roamer Calls Someplace Other than Home

When a roamer places a call to someplace other than his home area, he of course has to pay the roaming area's per-minute charges. He is also billed the long-distance rate between the MTSO he is using and the place he is calling.

A Roamer Gets a Long-Distance Call from Anywhere

Again, the roamer gets to pay the airtime charges between his phone and the MTSO he is using. Fortunately (for him), the caller pays the long-distance charges.

What it all boils down to is this: When you're roaming and call out, you pay all the airtime charges as well as whatever long-distance tolls you incur. If someone calls you, he pays only for the long-distance call between his phone and the MTSO you are using in your travels; you still pay for the air time you use.

FOR MORE INFORMATION

If you are unable to obtain roamer service automatically, most carriers have a standard number to call for information and help. It can only be accessed from your cellular phone, and it's free. Usually it is 711 or 611, sometimes preceded by an asterisk. You may want to call just to check current prices.

Some carriers will call you the first time you enter a city and your phone is on, because the MTSO will sense your phone's presence. They will welcome you as a roamer and answer any questions.

8 | Portable and Transportable Phones

To be able to place and receive telephone calls from your car may not be enough; you may need that convenience in a factory or office building, out at a construction site, or even on the bay in your boat. The miniaturization made possible by today's electronics technology makes your wish easy to realize.

TYPES OF PORTABLE PHONES

Portable cellular phones fall into three general categories: briefcase-size units, transportables, and handheld phones. Each type has something to recommend it.

Briefcase-Size Phones

The first portable cellular phones contained the transceiver/logic unit, handset, and power pack in a package that, when closed, looked like a briefcase. (These phones are no longer widely available.) The rugged shell protected the contents of the case from damage, making it well suited for use in remote, rough-and-ready situations. With a sturdy case even bouncing around in the back of a truck won't cause significant damage.

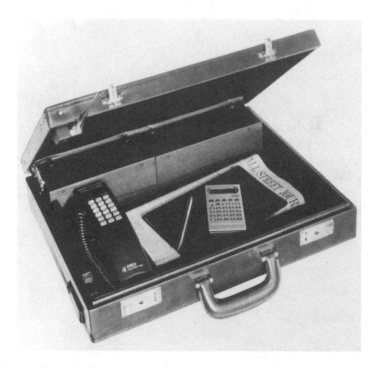

Briefcases offer a convenient way to transport all the components of your phone.

Inside the case are all the components of a cellular phone. Briefcase-size units are usually supplied with antennas, similar in appearance to the "rubber duckie" antennas used with Handie Talkies, that mount on a connector outside of the case. This connector can also be used to connect a vehicle-mounted antenna for road and highway use (or to get better range than might be provided by the flexible antenna).

This rugged case is well suited for use in remote, rough-and-ready situations.

The case also contains a twelve-volt rechargeable battery power supply in the form of nickel-cadmium cells or special lead-acid batteries. The care and feeding of these power packs is discussed later in this chapter. There is usually a meter or other indicator to monitor the condition of the battery pack, and there is also usually a power jack that allows it to use a vehicle's power supply. The cable that plugs into this jack frequently has a cigarette lighter adapter on its other end so it can draw power from that source.

The power output of these phones is generally higher than that of other, smaller portables. It is usually the maximum permitted, three watts, as contrasted to the handheld units' .6 watt (600 milliwatts). Consequently, the batteries of these phones are heftier, which contributes significantly to their weight—often twenty pounds or more.

Inside the pack are all the components of the phone.

With the increasing availability of transportables and portables, the briefcase phone has virtually disappeared. However, it has taken on a new life as a mobile work station, or
portable office, including a smaller full-power phone,
fax/data modem, and notebook personal computer instead.

Transportables

Many phones that can be mounted permanently in a vehicle
can also be mounted semipermanently, to be removed and
fitted with an adapter for portable use.

Typically this configuration consists of a bracket with a
handle. The bracket holds a rechargeable battery pack, and
the phone's transceiver/logic unit and handset can be
attached and removed quickly (see Figure 8.1). In portable
use a rubber duckie-type antenna is usually connected in
place of the cable that feeds to the vehicle-mounted
antenna. If the initial phone installation is done with this
dual use in mind, it is a simple matter to switch between in-
vehicle and portable use.

Also available from a number of manufacturers are brief-
case- or knapsack-type cases to convert mobile-mounted
units into transportable ones. The result is similar to those
described above.

Like the transmitters used in briefcase-size units, those
of these transportables are full-power units, outputting
three watts (as you might expect, since they are designed
for vehicular use). Their battery packs may not have the
capacity of those used in the larger units, sometimes permit-
ting only four hours of operation instead of eight or twelve.
A few manufacturers give you the choice of configuring
your portable phone with different-sized battery packs.
Those allowing extended operation cost—and weigh—
more.

If you use your portable phone both in your car and out
of it and plug it into the cigarette lighter receptacle to take
advantage of the car's larger battery, find out whether that
receptacle is wired into the same circuit as the ignition
switch. If it is, power will be available at the receptacle only

FIGURE 8.1
PARTS OF A TRANSPORTABLE PHONE

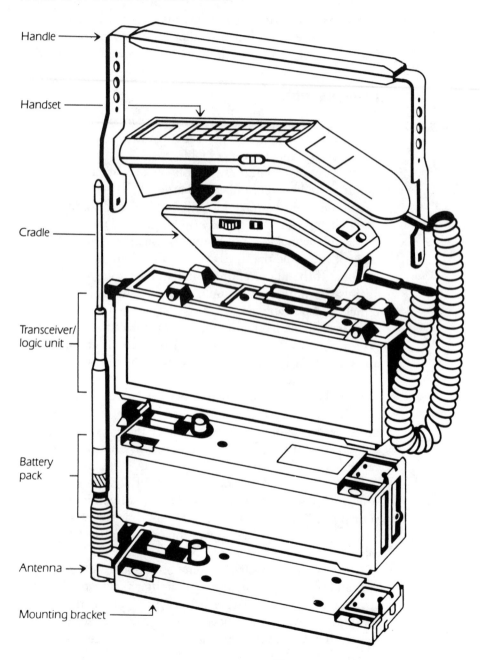

Handle

Handset

Cradle

Transceiver/
logic unit

Battery
pack

Antenna

Mounting bracket

With a transportable phone, you can keep in touch with your office from a remote job site.

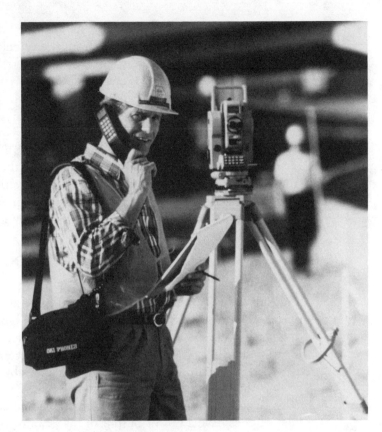

when the ignition is on. Your phone won't work and its batteries will not be charged if the ignition switch is off. If this is the case, don't expect to leave your phone in your garaged car overnight and find it fully charged in the morning.

A transportable phone of this sort and its battery pack weigh four pounds or perhaps a bit more, depending on the power pack. In addition to a carrying handle, most phones of this type also allow a shoulder strap to be attached, making carrying the unit for extended periods somewhat easier.

These transportables can also be used in their portable configuration in a vehicle. Unscrew the rubber duckie antenna from its connector, and attach the cable from a permanently mounted one in its place. As with the briefcase-size phones, an adapter allows use from the vehicle's own battery, usually via the cigarette lighter. As with the larger units, this adapter can also recharge the battery pack. Adapters may also be available to allow the phone to be taken indoors and operated from house current. These adapters can also be used for battery recharging.

Some adapters may not be powerful enough to both operate the phone and recharge its batteries at the same time. You may have to turn the phone off to use the battery-charging feature. Since most battery packs require between twelve and sixteen hours to be fully recharged, you can usually do this at night.

Handheld Phones

The smaller a phone is, the easier it is to carry around. The portable phones discussed so far free you from being tied to one place while making calls or waiting to receive them. Handheld cellular phones reduce this burden even further.

To get so much into such a small package—about the size of a large police-type Handie Talkie radio in most cases—certain sacrifices have to be made. While larger cellular phones have a power output of three watts, the handhelds typically have an output of .6 watt (600 milliwatts), or one-fifth that of the larger ones.

As with most things electronic, when sizes get smaller, prices get bigger. Many handheld cellular phones are priced in the $400–1,000 range, considerably more than most of their larger brethren. What you are paying for is not more features—these compact phones offer pretty much the same as the larger ones—but for putting all these features into such a small package.

Handheld phones use rechargeable nickel-cadmium (Nicad) or nickel-metal hydride (NiMH) batteries built into their cases, although some manufacturers permit use of regular alkaline batteries also. The phones are supplied with a charging cradle, which they sit in when not in use. These

This phone contains its own antenna that can be telescoped into its case.

cradles supply the power needed to recharge the batteries. Since the phones are all-in-one units, with the power pack, transceiver/logic unit, and handset/control unit combined in one package, they must usually be removed from their rechargers to be used.

These phones contain their own antennas, usually flex (rubber duckie) types, that can be telescoped into the phone body, out of sight where they can do no damage. In strong signal areas, a phone can be operated without extending its antenna, although getting it out in the open may improve its performance.

Although most portables are intended strictly for portable use, a few can also be used as part of a vehicular system. A cradle in the vehicle not only supplies power to the unit but also serves as the connecting point between a handset/

control unit and a power amplifier that increases the unit's output to a full three watts.

When most cellular systems were new and cell sites were far apart, portable service was noticeably spotty because of the portable's lower power. In-building and in-auto use were particular problems. Today, most cellular systems are built out to the point where no difference in performance can be detected between portable and full-power mobiles in most situations.

Thus, the majority of new buyers are opting for portables; they are more convenient, can be used in offices and other areas out of the car, and are easily carried along anywhere for business trips. However, they are more expensive, easy to misplace, and are a major source of irritation to their owners because most only allow about an hour of "talk time" between battery charges. A second battery is a must for most users.

Wireless Portable Phones

Another type of portable phone you may encounter is like the wireless phone you use at home—it is a portable phone for the phone in your car.

It uses a second transceiver connected to the cellular unit in your car and works just like home wireless phones. If you take the wireless handset with you (and leave the cellular phone in your car turned on), you are carrying your access to your cellular facilities.

The wireless phone allows you to place and receive calls within a radius of up to 1000 feet from your car.

HEALTH CONSIDERATIONS

From time to time rumors have surfaced that there may be some health risks to the use of cellular phones, especially portables, due to radiation from the tip of the antenna in close proximity to the head. While studies continue and more stringent standards on radiation are being considered, it should be pointed out that no studies have linked any cellular phones to health risks. Any assertions of health risks up to this point are merely rumors and uninformed accusations.

CELLULAR ETIQUETTE

The general availability of portable cellular telephones that can easily be carried in the purse, briefcase, or suit/shirt pocket has generated a number of new social situations inconceivable before the advent of the portable cellular phone.

Your desire to be in constant communication (or merely to impress people) can make you oblivious to your potential impingement on the rights of others when using a cellular phone. Try to be courteous in the following situations:

1. Don't put your cellular phone out on the table at a business meeting or at a restaurant; this sends a signal that you will interrupt the meeting or the meal with a phone call and that you value the phone call more than your present company.
2. Don't use a cellular phone at a meeting, even if at a break; excuse yourself to another room or the corridor.
3. Don't use the phone at a table at a restaurant; it is discourteous to other patrons as well as those at your table; excuse yourself to the lounge.
4. Don't let your portable phone ring at a theater, meeting, or conference, much less answer it and use it. Turn it off and use a pager, preferably with an inaudible signal. If you are expecting an important call, reschedule it.

POWER SUPPLIES

Whether you use it in your car or in your hand, a cellular phone operates from one type of rechargeable battery or another. The only exception to this rule is portable phones that use ac adapters to operate from house current. Some of these will automatically disconnect from their batteries when the adapter is plugged into them. A phone may operate from its own battery or from the battery of the vehicle in which it is being used.

The information that follows will help you extend the life of the batteries that keep your phone operating. And, if you depend only on your car battery to power your phone, this advice will help you keep it, too, in better shape.

Battery Types

Cellular phones rely on two types of rechargeable batteries: nickel-cadmium cells and lead-acid cells.

Nickel-Cadmium/Nickel-Metal Hydride Cells

These operate such household appliances as cordless vacuum cleaners, electric mixers, and conventional wireless phones. They are the only type found in handheld cellular phones.

Nickel-cadmium cells, sometimes referred to by their trade name, Nicads, have an unusual characteristic. They can develop a preference for being used in a certain way and remember that preference. If you use a phone powered by nickel-cadmium cells just a few times before recharging it, instead of running it until its batteries are nearly exhausted, it will begin to demand recharging well before it should be necessary. In other words, it will develop a preference for not holding a charge very long, not nearly as long as it was intended to.

To break in a set of nickel-cadmium cells properly, you should, *before you ever use them*, charge them for twenty-four hours. Then, use them hard, until their charge is nearly depleted. Then charge them up again for the minimum normal charging time, twelve to sixteen hours.

When you use your phone, try to recharge the batteries only when they need it; don't leave it in its charger when it doesn't really need it. Putting a full charge on the phone's batteries when they still have some power left in them will only lead to development of those memory characteristics that will cut into their working life.

The abysmal and short-lived performance of so many rental phones is, in many cases, due to their batteries having been mistreated in the charge-discharge cycle. Perhaps this is unavoidable in the lives they lead, but it doesn't have to happen to your phone.

Should you come across or develop a set of nickel-cadmium cells with memory problems, you can often restore them by running them through two or three complete charge-discharge cycles. Charge them up, run them almost all the way down, then charge them up again. Several repetitions of this will frequently resurrect nickel-cadmium cells that were thought to be dead.

Nickel-metal hydride (NiMH) batteries offer greater capacity for their size and are not subject to memory effects.

Lead-Acid Cells These are the same cells that make up your car's twelve-volt battery. They are not used in hand-held phones but may be found in some larger portables, particularly briefcase-size ones. Because these batteries are liable to be shaken up and even carried in some unlikely positions, the liquid acid electrolyte used in automobile batteries is replaced by a gelled or immobilized electrolyte, which is much safer.

A *gelled-electrolyte battery* contains an acid electrolyte that has been converted, through the addition of chemicals, into a gel that cannot leak or spill. In an *immobilized-electrolyte* battery, the liquid acid is contained in a spongelike material that keeps it from sloshing around, as well as preventing leaks and spills.

Lead-acid cells are just the opposite of nickel-cadmium ones in their recharging requirements. The latter should be used "until they drop" before being recharged, whereas lead-acid batteries should always be kept as fully charged as possible to extend their life. If they are allowed to run down too far, they will be unable to hold a charge for very long. (This is one reason people connect trickle chargers to the lead-acid batteries in cars and other vehicles that aren't going to be used—and charged—for a long period. It keeps them in top shape.) If your phone uses lead-acid cells, try to charge it whenever you can. This will extend the phone's battery life immeasurably.

Operating Life

The useful working life of a battery pack on a single charge is usually specified by the manufacturer. Be aware, however, that this period refers *not* to the length of time the phone is in use but simply to the time that is *on*. In other words, a battery pack may last six hours if it's used to power a phone waiting for a call, but it will be good for only an hour or so of actual connect time.

9 Options and Accessories

Cellular telephones have so many features built into them that it's difficult to think of anything more that could be added to them. Many of these features have been discussed in previous chapters, but we will explore them further here.

ADDITIONAL STANDARD FEATURES

Before we look at the options that can be added to cellular phones, let's look at a few more of the built-in features that are standard equipment.

Timers

Most cellular phones have built-in timers of one kind or another. The simplest kind tells you the length of the last call you made. This is of somewhat dubious value, since what you would really like to know is how long you've been talking during the current call. Some phones emit a beep at regular intervals, to help you keep track of how long you've been talking. Some timers can tell you how long you've been talking, how long you've used the phone (the timer can be reset to zero whenever you like), and the total number of calls you have made.

And, since a the existence of a timer implies that of a clock, some cellular phones also include one, complete with day and date information.

Hands-Free Operation

When you were learning to drive you were told to keep both hands on the wheel! When you started using a cellular phone, you discovered you had to use one of those hands to hold the phone. The handsets of cellular phones are *not* designed to be crooked between your neck and shoulder since holding your head at that awkward angle would make it extremely difficult to concentrate on the road.

Most manufacturers of cellular phones and a few outside suppliers offer hands-free accessories that make it unnecessary for you to hold the handset while carrying on a phone conversation. The simplest of these replaces the conventional handset with a headset similar to those used by telephone operators and astronauts, which contains an earphone and a small microphone on a boom in front of your

This hands-free micro-phone system attaches to the driver's seat belt.

mouth. These headsets can be a great convenience, espe-cially if you spend a lot of time on the phone. Be aware, however, that some states may prohibit their use in moving vehicles since they're considered to be in the same league as portable cassette-player headphones.

A more sophisticated hands-free arrangement consists of a speaker, which may be built into the phone's control head, and a small microphone, which clips onto your windshield's sun visor to pick up your end of the conversation (see Fig-ure 9.1). For privacy, removing the handset from its cradle restores it to operation and deactivates the external speaker and microphone.

At least one phone goes a step further and frees you from even having to dial a number. It features a speech recogni-tion system that understands simple spoken commands. The phone even prompts you, using its own voice. When you pick up the handset, it says, "Name, please," and waits for you to tell it who to call. You may respond, "Bob Jones," to which the phone will reply, "Dialing 817-555-

1234." If the line is busy, you can tell the phone to redial, and, when you dial a call manually, the phone speaks the digits as you enter them, eliminating at least one possible source of error.

Still another phone, while it isn't intelligent enough to understand what you say to it, does make it easier for you to dial. The keypad is separate from the phone's handset and can be mounted in a convenient and easily visible place, such as a car's dashboard. You mount it where you can see, reach, and dial it easily, sacrificing very little in the way of control and concentration.

FIGURE 9.1
ONE TYPE OF HANDS-FREE OPERATION

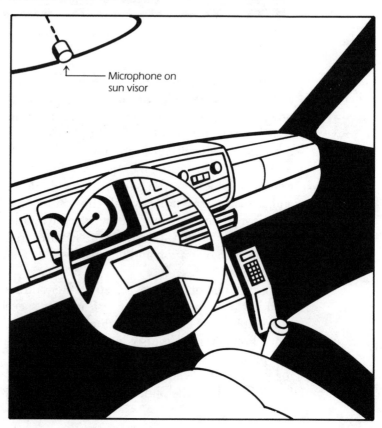

Adding voice recognition capability to cellular phone installations allows users to vocally enter telephone numbers.

(The issue of driving safely while using a mobile phone is extremely important. A number of the factors involved will be examined in Chapter 11.)

Other Built-In Features

Most cellular phones include provisions for a *horn alert*. This feature sounds your vehicle's horn when a call is received. If you spend a lot of time in the field, away from your phone but within earshot of your car, this will alert you that your phone is ringing.

Still another feature built into many cellular phones is an indicator that you received a call while you were away from the phone (but only if you left the phone turned on). When you return to the phone, you will find the word *CALL* showing on its display. The phone won't be able to tell you who called or when, but if you were expecting a call, you probably got it and can now return it. Some models actually prompt a caller to leave their number using their touch tone pad. So now your phone *can* tell you who called!

Finally, many phones include a dual or multiple NAM capability. This permits the user to register as a home subscriber on more than one system. For roamers who travel between two or three locations frequently, this may save them money.

ADDING PORTABILITY

After you've used a cellular phone in your car for a while, you will probably decide that you can benefit by having it with you at other times as well. Many manufacturers have designed their phones with both permanent installation and transportability in mind and offer options to make the changeover from one mode of operation to the other fairly simple.

Some phones—these are usually mounted up front, near the driver's position—can be removed from a vehicle with the twist of a couple of quick-release connectors and snapped into an adapter, complete with battery pack, antenna, and handle or shoulder strap.

Carrying cases can range from soft knapsack-type ones, which add portability to luxurious attache cases (that can also be used to carry papers and other materials) to rugged boxes intended to be tossed into the back of a truck or jeep and bounced around the countryside all day. The manufacturers of these cases usually offer a number of accessories to go with them. There are antennas of all sorts and, of course, your choice of battery packs, with rechargers of vari-

Knapsack-type carriers add portability to your phone.

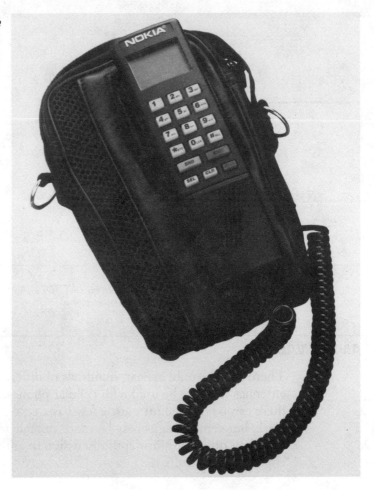

ous capacities, offering four, six, or eight hours (or even more) of use before they need recharging.

The prices of these accessories vary, depending on the ruggedness or luxuriousness of the carrying case and the capacity of the battery pack it contains. For a few hundred dollars, though, you can convert most cellular mobile phones to transportable use. With the continuing drop in the price of portable phones, you may want to consider buying a portable rather than converting an existing phone.

"... *Plus I want to buy 200 shares of CTM Systems and sell all of the Acme Chemical, then convert the* ..."

ANTENNAS

There are, it would appear, hundreds of different kinds of antennas that can be used with cellular phones. Actually, these can be divided into just a few basic types. All antennas work, but some designs may be more suitable to your situation than others. An in-depth discussion of antenna types appears in Appendix A.

DATA COMMUNICATIONS

Telephones are not the only devices that have recently gone portable. Computers have also become portable, with some offering all the power of desktop systems in packages little larger than a loose-leaf notebook.

One use to which these laptop computers are frequently put is communicating with other, larger computers—either to send information (perhaps a day's orders taken by a sales-

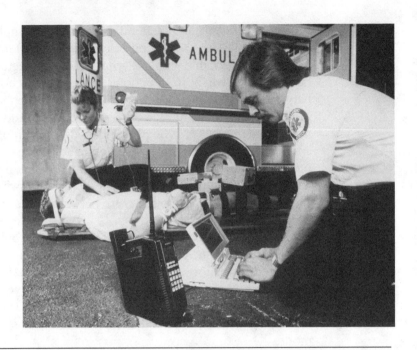

Transportable phone with data interface capability used with small laptop computer.

person working in the field) or to receive it (for example, to check on the status of an order already being filled). These transportable computers can also be used to write—and transmit—letters, memos and, reports and even to make airline reservations.

All this computer-to-computer communication takes place by telephone. Digital signals output by your computer are converted into audio tones by a device called a *modem* and sent to the distant computer over standard telephone lines, where they are converted back into digital form. Information from the remote computer is transmitted back to you in the same way.

Some cellular phones come with built-in modem connectors, and several manufacturers have designed modems specifically for cellular use. Many laptop computers even contain their own modems.

If your cellular phone has a data connector, simply plug your modem (with the computer connected to *it*) into it. You will need a cable with the appropriate connectors at both ends. A switch on the phone allows you to choose between voice or data communications. Cellular phone modems can also be used with most phones that do not have data connectors. The phone's handset is unplugged, and a cable from the modem (there are adapter cables for most makes of phones) is connected instead. The phone's handset is then plugged into the modem, and a switch on the modem allows you to alternate between voice and data transmission.

Cellular modems have advanced in speed capability as fast as their landline equivalents. Like landline modems, cellular modems can work at speeds in excess of 14,400 bits per second (bps) over regular phone lines. However, since cellular introduces several additional variables that can increase error rates over landlines, 1200–9600 bps is the mode.

There are several problems that can arise in cellular data communications, and cellular modems are designed to take them into account. The biggest obstacle to reliable data communications is signal dropouts. These sometimes occur even on conventional telephone lines and may also be encountered in cellular systems. In data communications, even an interruption of a couple of milliseconds—just a few thousandths of a second—can be disruptive. So can bursts of noise, which the computer may attempt to treat as data even though they are nothing of the sort. Even the infinitesimal, and usually unnoticeable, interruption that may take place as a call is handed off from one cell site to another can wreak havoc with data communicated by modem.

The errors that may be induced by signal fading, dropouts, or noise cannot be eliminated, but they can be corrected. There are two ways in which this can be done.

The first method uses an error correction scheme built into the modem. As the modem sends out data, it also periodically transmits a number that it has calculated based on the values of the numerical representation of that data. (In computers, everything is treated as numbers. It's only when a computer has to present information to you that it's translated into terms, such as words or graphics, that you can understand.) The modem at the receiving end performs the same type of calculation on the data it receives. If the figure it arrives at disagrees with the one that has been sent with the data, it knows that an error has crept in and requests that the information be transmitted again. The chances are that the second time it will come through correctly.

This error correction by modem is particularly useful when information is being sent from the keyboard of one computer to another. It can't fix typing errors, but it does ensure that what gets typed is received correctly at the other end.

Information can also be sent from computer to computer directly, without entering it from the keyboard. Files (collections of data, including documents prepared using a word processing program) can be created and stored on floppy disks and then transmitted from one computer to another all at once, which is much faster than they could be transmitted if they had to be typed in character by character at the time of transmission. Sometimes this is the only way certain information can be sent; at other times it simply provides the means to cut down on connect time and keep airtime charges to a minimum.

While error-correcting modems can be used for these direct-from-disk data transfers, virtually all communications programs—the programs that allow computers to send information by modem—also include some form of error correction for this mode of operation. The systems embodied in these programs work in much the same way as those used by the modems. They derive a *checksum* or *CRC* figure from the data being sent and use it to ascertain the validity of those data when they are received. If the transmitted and received figures differ, the block of data in question is retransmitted until they agree. Of course, programs using the same type of error checking must be used at both ends of the communications link.

Ordinary (deskbound) modems are quick to hang up when they no longer hear the signal from the modem at the other end of the line. When one modem hangs up, so does the other. With dropouts being a fact of cellular life, this could present a problem. Consequently, modems intended for cellular phone use are more tolerant of signal loss. They have a longer time-out period, and some even synthesize a connect indication when the real one temporarily disappears.

Needless to say, it would be extremely unwise to attempt to use a computer and a cellular phone while driving. More about this is discussed in Chapter 11, "Safety and Security."

SECURITY

There are two types of security applicable to cellular phones. The first deals with the theft or unauthorized use of the phone equipment. The other concerns itself with the possibility that conversations or computer data sent over a cellular phone link may be intercepted by parties other than the one they are intended for.

Both of these issues and some of the security devices available are discussed in Chapter 11.

FAX

Because of the popularity of facsimile (fax), portable fax machines are available from many manufacturers. Most are specially designed to interface with cellular phones. While the use of portable fax machines is still a specialty, like portable computing it is becoming an important part of the "mobile office."

With a portable fax machine you can send and receive documents via your cellular phone.

Some cellular fax machines are intended to be used with an existing phone; others come with their own cellular phones built in.

VOICE MESSAGING

Another recent development is voice messaging, which works sort of like a phone answering machine without the answering machine. Sure enough, there are now voice messaging devices for cellular phones. Voice messaging is usually offered as a special service by the cellular carrier, through a specialized computer attached to the MTSO. In conjunction with call forwarding or no-answer transfer, it will answer your phone when you don't care to or are not available. Some of these services can call you or page you to tell you that you have an important message waiting, and permit you to exchange voice messages with other users.

Finally, some voice messaging services provided by cellular carriers will soon allow you to receive, store, and send fax messages. This permits you to have faxes sent to you at any time, re-route them to a fax machine convenient to your next destination, or print them all on your portable fax at a convenient time. Or, you can send a fax to your voice mailbox, and then "broadcast" it to an entire group without having to send it to each person individually.

Suffice it to say that, if a device exists to enhance the operation of home- and office-bound phones, there is probably also a version of it—or soon will be—for cellular phone use as well.

10 Dealing with Operational Difficulties

Cellular telephones are rugged and reliable. They have to be since their very nature dictates that they will be banged around and subjected to extremes of heat and cold in the trunks of automobiles and other places. They rarely fail. Most of the difficulties with your cellular phone will arise from the conditions under which it is used and, possibly, from your initial unfamiliarity with the equipment and with cellular systems in general. This chapter will help you to understand some of the difficulties you may encounter in using your cellular phone and, when possible, will show you what you can do about them.

BAD CONNECTIONS

As in any phone system, once in a while you will get a bad connection with your cellular phone. This is particularly true in fringe areas—near the edge of a company's area of service or where the terrain interferes with the system's radio signals. Often it is possible to continue a conversation under poor conditions, but sometimes things get so bad that you are automatically disconnected.

Disconnects occur because of signal dropouts—either your phone loses the signal from the cell site or the cell site loses the signal from your phone. These two things do not always happen simultaneously—sometimes you will be able to hear the other party but he won't be able to hear you or vice versa. Sometimes there is a way to tell when a disconnect is going to happen, allowing you to exit gracefully from a call before you are cut off.

All phone systems—landline and cellular—provide some sort of audio feedback. That is, when you speak, you can hear a portion of the electrical signal in the earpiece of your phone. You're usually not conscious of this little voice whispering in your ear, but you always hear it, and it reassures you that your phone is operating properly.

When you don't hear this feedback signal, you know you're in trouble. Those times when you pick up a phone and it is dead, that deadness you hear is due to the absence of the feedback signal. Try this experiment on your home phone. Pick up the handset, put it to your ear, and blow into the mouthpiece. You'll hear yourself blowing in your ear, as it were. Now unplug the phone and do it again. You'll hear the difference immediately—no feedback.

Sometimes when you're speaking on a cellular phone you'll notice a momentary loss of feedback as you talk. This indicates either that the cell site lost your signal for a moment or that you lost its signal.

Cellular phones are designed to take these momentary dropouts into account and will not disconnect unless a dropout exceeds a certain period. However, if you experience a number of dropouts, one after the other, be prepared to be disconnected. Multiple dropouts are a sign that you're in an area of poor signal reception or transmission and that there may be more, one of which may be long enough to cause the system to disconnect you, thinking you're no longer there. Should you experience these dropouts, tell the party at the other end of the line that you may be disconnected from him abruptly and that you'll call him back if this happens.

SIGNAL DROPOUTS AND DEAD SPOTS

Signal dropouts are annoying. They can ruin conversations by causing you to lose words, by having to repeat words, and by being disconnected. *Dead spots* are regions where your cellular phone doesn't work at all, and *dropouts* occur when you run into a series of dead spots. There isn't much you can do about these when you encounter them, but there are steps you can take to avoid them.

What causes dropouts and dead spots? The answer lies largely in the behavior of radio waves at the frequencies used by cellular phones. At these frequencies—almost a billion cycles per second—radio waves start to act a bit like light waves. They travel in straight lines, can be weakened by water in the air (which is why UHF TV signals aren't as strong on rainy days), and are easily reflected by man-made and natural objects (the same principle that makes radar work).

For reasons that are not entirely clear, some areas are just naturally dead to radio waves. This may be due to the terrain or to disturbances in the signal path caused by objects many miles away; whatever the cause, some areas are just not good for reception or transmission at certain frequencies.

Dead spots are frequently encountered in mountainous or hilly terrain and sometimes in heavily wooded areas. But they are not restricted to rural areas (see Figure 10.1). There are many dead spots in cities. These have two causes.

The first is simply the blockage of signals by buildings between your phone and the cell site. Although the site antennas are located as strategically as possible for optimum coverage, there is bound to be something in the way somewhere.

Dead spots also occur because of *multipath reception* (or just *multipath*), which happens when two or more signals interfere with one another. The signals may be different or they may have the same source (see Figure 10.2).

One example of multipath is the "ghosts" on your TV screen. The original TV signal arrives at your antenna from two or more different directions. One signal comes directly from the TV station's transmitter, but the others (there are usually more than one) are portions of that same signal that have been reflected from buildings in your vicinity. Because the path traveled by the reflected TV or radio signals is slightly longer than that traveled by the direct signal, the reflected ones arrive at your TV antenna slightly later. Even at the speed of light, your TV set discerns the time

FIGURE 10.1
DEAD SPOTS CAN OCCUR IN MOUNTAINOUS TERRAIN

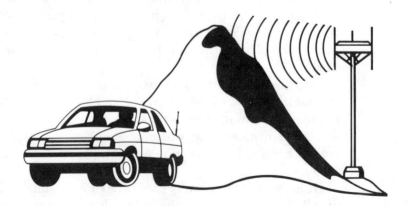

delay and displays the picture carried by the reflected signal a minuscule fraction of a second later than that carried by the direct one, causing the ghost on the screen.

Cellular telephone signals can be reflected in the same way and, since the direct and reflected waves can mix and cancel out one another, communications can thus be seriously affected (see Figure 10.3). When the peak of one radio wave mixes with the trough of another, the result is effectively zero—the combination results in no signal at all or a badly distorted one. If you are at the spot where this happens, you will be in a dead spot. Depending on conditions, dead spots can be small or huge.

FIGURE 10.2
MULTIPATH REFLECTION

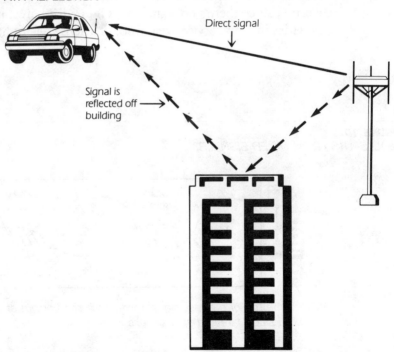

Direct signal

Signal is reflected off → building

Multipath reflected waves can mix with each other (and with direct signals), resulting in dead spots or badly distorted reception.

Dropouts happen when you pass through a number of dead spots, one right after the other. The conditions that give rise to one dead spot frequently give rise to many.

What can you do about dead spots or dropout areas? Not much. If you routinely travel the same route, try changing your path by a block or so or try not to use your phone when you're in a trouble spot. You may find, especially in the case of multipath, that simply moving a foot or two will restore signals both ways. (That's why you can make TV ghosts shift or even disappear by adjusting an indoor antenna slightly.) If this doesn't work, you'll just have to wait until you're in a better spot.

Dead spots are frequently responsible for causing a phone's NO SERVICE indicator to light. It comes on not because there's no cellular service in the city you're in but because, since there's no signal in its immediate area, it *thinks* it's in a no-service area. And, in a sense, it's right.

FIGURE 10.3
DEAD SPOTS FROM MIXED SIGNALS

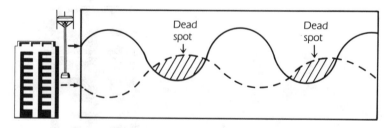

Where the top of a direct signal mixes with the bottom of a reflected signal, a dead spot occurs.

There is, however, a bright side to the multipath phe-
nomenon. Sometimes, even if the direct signal is blocked,
you may be able to use your phone via a reflected one (see
Figure 10.4). There is no rule to guide you in this, but
you'll often find that you can make and receive calls from
the most unlikely places.

FIGURE 10.4
REFLECTED SIGNALS

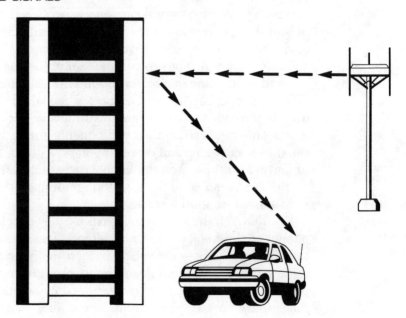

*Even if a direct signal is blocked, your phone may be able to use
a reflected signal.*

OTHER INTERMITTENT EFFECTS

When placing a call from an area of marginal service, your phone's IN USE indicator may come on and then, after a few seconds, be replaced by the NO SERVICE one. This happens because, when the phone makes contact with the cell site, it (or the cell site) decides that the signals are not good enough to maintain a connection and disconnects you. There's nothing to do for this but move elsewhere.

Similarly, you may find that you even though you can carry on conversations when passing through certain areas, you can't place calls from them. This is because the thresholds of the cellular system are set to stop a call being initiated when the chances are that it won't be successful but to allow an in-progress one to continue for as long as possible.

Dead spots come in all sizes and intensities of deadness. "Picket fencing" or "flutter" happens when you drive quickly through alternate zones of strong and weak signals, causing the signal strength at your antenna (or at the cell site's) to fluctuate up and down. If you slow down, the rate of flutter also slows. You can usually carry on a conversation under flutter conditions, suffering nothing other than a little annoyance until it passes.

Even though the FM radio signals used by cellular phones are immune to most interference, you may still occasionally experience—particularly in metropolitan areas—some static. This is often due to interference from nearby powerful transmitters, such as those used by broadcast stations. (New York City's infamous Intermod Alley is located directly beneath the Empire State Building, which bristles with broadcast and other commercial service antennas.) You can't do anything about this but get out of their way, but it's reassuring to know that the difficulty is not being caused by your phone.

Static may also interfere with your conversation when you are near the limit of a region's coverage with no adjacent cell site to be handed off to. As the signal level drops, outside noise—normally covered up by the cell's signal—may sneak in. Should you encounter this, tell the person with whom you are talking that you may be suddenly disconnected if the cell-site signal drops so low as to be useless.

Indoors is one place you might not expect to have difficulty in using your cellular phone. Yet, in certain buildings—even if they're right underneath the cell site—you may find your NO SERVICE indicator constantly lit, or find yourself disconnected when you talk while carrying your portable phone from place to place.

This is because the steel used in high-rise buildings for beams and girders, and for the reinforcing rods and mesh used to strengthen concrete, works as a radio wave shield. Notice how the performance of broadcast band AM and FM radios falls off in certain parts of office and apartment buildings. Reception may be good near the outside walls, especially by windows, but it becomes more difficult the deeper into the building you get. The effect is the same on cellular phones. (It is these same materials, by the way, that *reflect* radio signals and are responsible for the multipath interference discussed earlier.)

If you must use your cellular phone in a metal-frame building (wood-frame construction won't faze it), try it near a window—preferably one facing in the direction of your cell site. Or try positioning your phone near or under an air conditioning duct. The duct acts as a pipe for high-frequency radio waves and may convey your signal into the open. Finally, because of the peculiar characteristics of high-frequency radio waves, just moving your phone six inches or so in one direction or another may help you to establish a connection or clear up a noisy one.

MECHANICAL FAILURE

A time may come when your cellular phone refuses to work. Before you blame the manufacturer or the dealer who sold it to you, try to find the source of the problem yourself. It may be—and usually is—something that's completely unrelated to the phone proper, and you can often fix it yourself with a minimum of fuss.

Power Problems

When you turn your cellular phone on and nothing happens—no lights, no reassuring beeps as the unit checks itself out—it's probably a lack of power. This usually happens when portable phones are repeatedly removed from and reinstalled in vehicles.

Check your connections. Maybe it's something as simple as a power cable that you forgot to reconnect. If your phone is powered from the cigarette lighter receptacle of your car, make sure the ignition switch is turned on. In many cars, power to the lighter receptacle is controlled by this switch, and, if the ignition is off, the phone won't work and it's batteries won't charge.

Another power problem is blown fuses. A blown fuse does not always mean that something is wrong with your phone—sometimes these things just happen. Know where the fuses that protect your cellular phone are located, and check them if the unit does not come on.

If you've been using your phone from its own battery pack, be sure that there's a usable charge on the batteries—maybe you left the phone on all night. And, of course, make sure all the connectors are firmly attached to their sockets and plugs.

Antennas

If you find you're getting a NO SIGNAL indication when you know you're in a good signal area or if the operation of your phone seems to be deteriorating, the problem may be in your antenna circuit.

First, make sure the antenna is still there. Many antennas are easily removable so you can stow them in the trunk of your car. If you have this kind of antenna, make sure to screw it on tight when you reinstall it.

Or maybe it's the other end of the antenna system that is the problem. Make sure that the cable coming from the antenna is connected to the phone. If it isn't, there will be no signal, and you will be without service.

If you connect and disconnect the cable between the antenna and your phone's transceiver frequently, a break may develop where the cable connector is attached.

A-B Switch

Although the A-B switches of most phones can be programmed for a priority mode—that is, they will always look first for an A or B carrier, depending on how they are programmed—it is better to lock into the service you normally use at home and when you roam, switch it to the appropriate service. This will ensure that—should you enter a spotty region within your home area—your phone will not accidently lock onto the service you don't subscribe to, which would get you no service or a recording telling you to call that service's business office.

If your phone is locked into the A or B mode when no A or B service is provided where you are at the moment, you will get a NO SERVICE indication. Try switching to the alternate service, although if no reciprocal roaming agreement exists and if you did not preregister your phone with the foreign system, you will not be able to receive or place calls.

Unless you know which service you will be using in a roaming area or unless you must lock in one service or the other (if a roaming area has two carriers and you or your home service company have an agreement with only one of them), set up your phone to switch automatically between the two. If for some reason—perhaps the lack of a roaming agreement—the service your phone selects is incompatible, you will find out when you try to use it. You can then program your phone to use the alternate service, if it's available.

Finally, if your phone lights but you can't dial out, be certain that the LOCK indicator is extinguished. The phone's LOCK feature, discussed in the next chapter, prevents people from using your phone without your knowledge or permission. If you forget to unlock it, *you* won't be able to use it either.

ELECTRONIC FAILURE

Even the most reliable equipment may suffer a component failure. The conditions under which some cellular phones are used—lots of bouncing around and the high temperatures in the trunk of a car where most of the electronics may be located—are those most detrimental to hardware longevity.

At some time your phone may start to act up or refuse to act at all as a result of an internal electronic failure. Unless you have the training and lots of expensive test and alignment equipment to back you up, there is little you can do to correct such a failure.

If, after making sure all the phone's connections are secure and checking for cable integrity and the like, you decide that whatever is ailing your phone is nothing you can take care of yourself, return the unit to your dealer or to another source of reliable service.

A well-equipped service shop will check out your phone, isolate the problem, and correct it in a matter of a day or two. Most cellular phones are designed on a modular basis, with various parts of their circuitry situated on individually replaceable circuit boards. Once a problem has been located, a defective module can be removed and a good one installed.

In the unlikely event that your phone has to go into the shop, if you have call forwarding—which allows calls to your cellular number to be switched automatically to your home or office—call your cellular service and have it turned on (you can't do it yourself if your phone is dead) until you get your phone back.

11 | Safety and Security

This chapter concerns safety and security—*safety* in the sense of maintaining your usual standard of driving while using a cellular phone, and in safeguarding your phone equipment and *security* in the sense of guarding the privacy of your conversations.

SAFETY

Anything that distracts you from driving can be a hazard. A cellular phone, unless it's used with care, can be just such a distraction. Using a phone at the same time you're trying to concentrate on the road and traffic conditions can endanger both you and other drivers.

For this reason, try not to dial a number while you're moving. Dialing takes a lot more attention than you think; your mind, eyes, and hands are diverted from controlling your vehicle. The designers of cellular phones have included features to make the equipment safer; take full advantage of what they have provided.

One thing that makes cellular phones smart is their memories. Some phones can store thirty or more numbers, and even the simpler ones can hold a dozen or more. Recalling these numbers requires only the pressing of a couple of keys and then pressing the SEND key to set the dialing process in motion. Even so, your attention can waver during the few seconds it takes.

If you must make a keypad entry while in motion, leave the handset in its cradle. This position, in the case of a permanent or semipermanent installation, is more or less in your line of sight with the road. Leaving the handset in place until your call goes through ensures that you have one hand free for controlling the steering wheel.

Some of the phones and add-ons mentioned in Chapter 9, "Options and Accessories," make driving while using the phone less of a risk to yourself and others. These safety-oriented devices include phones that place their keypads directly in your line of sight—on the dashboard, for example—and make it unnecessary to pick up the handset until a connection is established. Some have speech recognition capabilities and allow hands-free dialing and conversation; others have separate microphones and speakers that do away with the necessity for using a handset unless privacy is desired.

Be especially careful if you're using a portable phone that's on the seat next to you. Looking over at its handset to verify that you're pressing the right keys takes your eyes from the road completely; you could forget for a moment that you're guiding a couple of tons of iron and steel at close to a mile a minute. Unless you're parked, dial these phones with the handset in front of you, not in its cradle on the seat alongside. You've almost certainly encountered, at some time in your driving career, drivers who were so completely engaged in conversation with the passenger sitting next to them that they were oblivious to everything else on the road. Bear in mind that this could be *you* when you're using your phone.

Without being aware of it, you may start paying more attention to what's being said or to what you're going to say than to the traffic or road conditions you're in the midst of. It's bad enough to miss your turnoff because your mind was

Don't let your mind or attention wander. If things get busy on the road, hang up and call back later.

on something else; it will be worse if you turn *into* something you didn't notice was there.

Don't let your mind or attention wander. Make safe driving your priority. If things get busy on the road, hang up and continue your call later.

Protecting Your Phone

A cellular phone is a valuable piece of equipment, and the antenna on your car can signal thieves that there is something inside worth a lot of money. It's simple to prevent them (or anyone else) from using the phone once they get their hands on it—just notify your cellular service—but why give them the opportunity to steal it in the first place?

If you use a self-contained or transportable phone, take it with you when you leave the car or at least disconnect it and stow it out of sight in the trunk or elsewhere. If the phone is permanently mounted, detach the antenna—many antenna styles are designed for quick removal—and store it out of sight. This, at least, will not draw so much attention to your car.

To discourage unauthorized use, most cellular phones come with a lock feature. By pressing the LOCK key on the keypad, the phone is set to accept incoming calls but will not permit dialing out. To unlock it, you must enter a code consisting of several digits known only to you. Some phones have two locking systems—one that is stored permanently in the phone's NAM and one that can be changed at any time by someone who knows how (you generally need to know at least the NAM unlock code to access the other).

There are a number of automotive burglar alarms on the market, and, if you store or use your car in a high-risk area, you probably should have one installed. Sometimes the sheer racket these devices make when tripped is enough to scare would-be thieves away.

SECURITY

Security as it applies to cellular phones is probably over-played. Certainly, when a telephone conversation, which is protected by privacy laws, goes over the air—as is the case with a cellular phone link—there is reason to be concerned. However, there is probably much less cause for concern than you may have been led to believe.

Early in this book we noted that some parts of the radio frequency spectrum used by cellular phones coincide with the upper reaches of the UHF TV band. Some high-end scanners may also cover these frequencies. The question is, how easy is it to eavesdrop on a cellular phone conversation—putting aside for the moment the legality of the matter—with this equipment?

The answer is, not very. The way cellular phones operate makes their signals difficult to locate and even more difficult to track. To eavesdrop on a conversation, you have to know two things: *when* it is going to take place and *where*. And, while the eavesdropper may have some idea when the words he is waiting to hear may be uttered, he has no idea where, among the more than 832 channels assigned to cellular telephony, those words are going to show up.

It's not like tapping into an ordinary telephone line and then sitting back and waiting or listening to the playback of a tape recorder. You must be at the right place at the right time, which is virtually impossible given the way cellular phones work.

The frequency pair, or channel, on which a conversation will begin is determined automatically by a cellular phone system's equipment according to the conditions that prevail at that instant. The location of the vehicle containing the mobile phone determines which cell site (of many) will be used, and each cell site has assigned to it a set of frequencies that differ from those used by adjoining cells. Which frequencies within a cell will be chosen for a particular conversation depends on the ones that are free when the call is made. Further, if a cell site has been split (see Chapter 2), the chances are that the new cells are served by directional antennas—that is, they concentrate their signals in a particular direction. A would-be eavesdropper on the wrong side of the antenna has little hope for success.

In addition to the initial problem of finding the correct frequency pair, cellular telephony adds the complication of frequency changes when a handoff is performed. Highly secure government and industrial radio communications use a similar technique (called *diversity transmission* and *reception*) to scatter a confidential conversation all over the radio frequency spectrum.

As the vehicle with a cellular phone moves out of one cell site and into another—usually a matter of only a few miles, no more than ten minutes under conditions where traffic is moving freely—the responsibilities for the radio link are transferred to that new cell site. And, since adjacent cell sites use different sets of frequencies to avoid interference with one another, the frequencies the conversation is transferred to will differ from those under which it was initiated. Again, frequency selection is done automatically, and there is no telling where the conversation will show up.

In theory, it should be possible to build a device that would emulate the receiver portion of a cellular phone and, once locked onto the correct control frequency (also a matter of luck), that would track at least one end of a call from frequency to frequency. In reality, such a device is extremely impractical.

Security Devices

Despite the extreme unlikelihood of anyone's coming across—and being able to track—your cellular telephone conversations, you may feel you need some measures to prevent your privacy from being compromised.

The best way to keep secrets from leaking is not to discuss them. When you are discussing matters of a sensitive nature on your cellular phone or when you think they may be mentioned, remind those at the other end that they are participating in a cellular phone conversation, a portion of which is going out over the air. Reminding them that their

conversation potentially open to public ears can prevent
indiscretions.

If you *must* talk about private matters, there are devices
to ensure that they stay that way. The first is a clamp-on
unit that you attach to your phone's handset. This small,
lightweight unit is powered by a self-contained battery and
can be used with almost any phone. It works on the principle of *audio inversion*, intercepting the sounds that form
words and changing their characteristics so they are unintel-
ligible without a reinverting device. The characteristics of
the audio inversion process can be modified by changing
the settings on a small switch in the voice scrambler; there
are usually tens of thousands of combinations. Only the
same combination set on an identical unit attached to the
phone at the other end of the conversation will produce an
accurate reproduction of the original speech.

This type of scrambler is, by current standards, a relatively unsophisticated device. Still, given the already built-
in safeguards against eavesdropping that cellular phones
provide, it should afford you all the extra protection you
feel you require. There are, however, more-elaborate protection devices available. These scramblers use digital techniques and complex encryption/decryption schemes to provide the utmost in privacy. Some can handle both speech
and computer data. These scramblers are usually connected
between a phone's handset and its transmitter/logic unit, in
much the same way as the cellular modems described earlier.

One such scrambling device divides small portions of
speech into even smaller "slices," and then "shuffles" them
into unintelligibility. A descrambler at the receiving end of
the communications link (ideally, both ends have encryp-
tion and decryption devices) reshuffles the voice slices
according to a complex decryption key and reproduces the
original voice with all the fidelity a cellular phone system is
capable of.

Devices of this sort are usually owned by the parties
using them (or by the companies they use them for). The
information they transmit is scrambled over the entire path
between one phone and the other. Some phone systems,

however, may offer a service that requires only the mobile cellular unit to have scrambling/descrambling equipment. The information transmitted, be it voice or computer data, is sent in encrypted form over only the portion of the phone link that uses radio. Once it arrives safely at the MTSO, it is decrypted by on-site equipment and completes its journey over ordinary landline in unencrypted form. No conversion device is required at the receiving phone, which means that a sensitive call can be made over a cellular link to anywhere.

The process is two-way; what comes from the office- or home-bound phone is encrypted at the MTSO and de- crypted by the equipment associated with the cellular one. The cellular service you subscribe to can tell you whether it offers this or a similar protection scheme. At least one state (California) is considering laws to make this type of scrambling mandatory, to protect the interests of users of cellular phones.

12 | The Future

The technology that permits us to put telephones in our cars and to carry them around with us will continue to evolve and provide us with new and improved services. Because there is no need for poles, conduits, or miles of expensive cable, cellular telephones can be located, either permanently or temporarily, wherever they are needed. This versatility gives cellular the crucial economic edge that ensures the rapid growth of this new industry. In this chapter we will look into the future of cellular telephones and their applications.

URBAN APPLICATIONS

Even in a major city such as New York, there are many areas where it is impractical to provide conventional telephone service. Parks and other recreation areas, such as beaches, would require an extraordinary outlay of funds to be provided with the telephone coverage they should have.

With cellular phone stations, phones can be located anywhere within range of a cell site. These may be permanent installations or, in the case of special events like a Fourth of July celebration or a park concert, may be moved around to different areas as they are needed.

Usually, these phone stations will not resemble telephones; rather, they will look like wireless police or fire call boxes. Should an emergency arise, all a civic-minded citizen has to do is push a button to be connected with an assistance center.

Cellular phones are ideal for this type of application because each phone's NAM contains identifying information unique to that phone. An assistance operator does not need to ask, "Where's the fire?" Using a computer to match a phone's ESN with its physical location, the operator can automatically pinpoint the emergency area. It is even possible to route the emergency call directly to the closest fire or police station.

On the highway, cellular phones are invaluable. Just picking up
this phone automatically connects you to an assist center that
knows where you are.

RURAL APPLICATIONS

There are many parts of the country where there is only
sparse phone service or no service at all. Such areas include
long stretches of some highways running through areas of
low population density, and national and state parks.
Despite the obvious advantages of being able to summon

emergency aid from such areas, it has so far been too impractical or expensive.

Cellular phones change all this. Without the need for cables to a switching office, rugged phones can be placed wherever they will be needed and left untended except for routine maintenance checks. In many parts of the country you don't even need electricity, as the phones can be powered by self-contained battery packs, and recharged during the day by solar cells.

TRAVEL

You can't use your cellular phone when you fly. One reason is that the signal it emits might interfere with the plane's own communications equipment and navigation gear. Also, from high in the air you would almost certainly be within range of a number of cell sites simultaneously, causing contention and confusion over which cell site—or which cellular system—was going to handle your call. Cellular phones, however, are currently being tested on some rail lines and, in the state of Washington, can be found aboard at least one hydrofoil commuter ferry.

TECHNOLOGY

Cellular phones, like all things electronic, are getting smaller. Although handsets won't shrink much (there is a limit to how small a keypad can be and be used by normal-sized fingers), what will change is the size of the phone's transceiver/logic unit. Already, pocket-sized handheld units demonstrate the extent to which miniaturization is possible, and further reductions are sure to come. We'll soon see phones in a holster, and maybe we'll finally see the commercial equivalent of Dick Tracy's wrist radio.

As phone sizes shrink, so will phone prices. Initially you pay more for a smaller, lighter version of what you already have. But increased production, as well as simplified designs and increased competition, will reduce the prices of many types of cellular phones.

The most significant hardware advance we'll see in the immediate future that will be especially applicable to hand-held phones will be in battery technology. A limiting factor in the usefulness of current handheld and transportable phones is the life of their batteries, the amount of space those batteries occupy, and the weight they add.

Energy storage devices (batteries) currently under development, and which should become commercially available in the next four or five years, will have working lives of perhaps 40 to 50 percent more per charge than the rechargeable devices available today. This means that from a battery

A pocket-sized cellular phone

pack the same size as the present ones, there will be an additional two to four hours of use. And, if size is a factor, these new power sources will make possible smaller and lighter phones with a life per charge roughly equal to that of today's.

CELLULAR SERVICE

The biggest and most obvious changes in cellular telephony over the next few years will be not in the phones themselves but in their service.

To begin with, the cost of service will come down. Since every city now has two carriers and perhaps some resellers, competition will lead to reduced rates and to package deals offering substantial overall reductions in cost.

There will also be more-reliable service. As the needs of primary and secondary cellular markets become satisfied, carriers will turn their attention to parts of the country they had hitherto ignored. You will, in a few years, be able to travel the length and breadth of the country and not be out of the range of cellular service.

Also, areas that already have cellular service will receive better service. As initial needs are met, existing cells will be split and new ones added to cover bothersome dead spots. Repeater transceivers that do nothing but relay signals from existing cell sites will also be added to fill the gaps, both in urban and in rural areas.

Cellular service will become easier to use because of the standardization agreements among carriers. Computerized roaming clearing houses will simplify the billing of roamers, and, as a result, the price of roaming will fall to levels comparable to your home phone bill. All billings will automatically be made to your home account. A uniform dialing method will be agreed on, and roaming agreements will proliferate. It will rarely be necessary for you to make your roaming arrangements in advance. You will automatically be logged onto any system you may use, and using it will be—with the exception of remembering to add the area code—as simple as placing or receiving a call from your own home service area.

DIGITAL CELLULAR

New technology will permit the radio portion of cellular to become digital. From the carrier's perspective, digital technology will allow more channels than analog. From the user's perspective, digital transmission will provide several advantages. First, digital will make it even more difficult to intercept conversations, thus providing even greater privacy of communication. But more importantly, digital provides clearer audio and more consistent communication. While the actual audio fidelity may not be much better, the background noise level will be reduced, and static, interference, and other competing noises which occasionally occur in the transmission should virtually be eliminated.

Digital cellular may already be available in your area. In order to make the transition, cellular phones that are capable of both standard analog, as well as the new digital transmission, are used. These are commonly called "dual-mode" cellular phones and are available in all types: portable, transportable, and mobile. While analog phones are not going to become obsolete soon, investigate these phones when you consider buying. In order to accelerate the transition to digital, carriers may have special offers on both phones and service that will minimize the cost of having the latest technology available to you.

Digital has its negative considerations also. There are two competing, incompatible standards for digital cellular. Time Division Multiple Access, or TDMA, has been accepted as a standard by the industry and was available first. This technology causes different conversations to take up different time slots on the same channel.

A competing technology, Code Division Multiple Access, or CDMA, "tags" digital pieces of the conversation with a code and sends them over a broad range of frequencies, to be reassembled at the other end based on the correct code. This permits even greater capacity than TDMA. However, CDMA is not yet accepted as a standard. Both technologies give the same advantages to users.

Thus, a dual mode digital-capable cellular phone that uses TDMA will not work in digital mode in an area that adopts CDMA. Some argue that a digital phone purchase should be put off until the standards are finalized; but because analog will be around, it will provide a lowest common denominator of nationwide service availability for a long time to come. If digital cellular is rapidly implemented, the prices of digital-capable cellular phones will soon be low enough for all users to afford to change to any new standards.

PERSONAL COMMUNICATIONS SERVICE (PCS)

New services have been proposed that include portable service in limited areas, or with no handoff capability. On the other hand, they may offer digital technology and user availability at a single number across the country. These services are still to be fully defined and include a whole family of data communications services as well. The Federal Communications Commission has already proposed frequencies for such services.

APPENDIX
A

Antennas

The antenna is vital to the performance of a cellular phone. Without it, be it nothing more than just a few inches of stiff wire, no signal could get into or out of a cellular unit.

Antennas for cellular phones come in all shapes and sizes (see Figure A.1). Some are intended for permanent mounting, some for easy removal. Some are intended for vehicular use, others to be attached to a portable phone. Some are designed to be mounted on metal, some on glass. There are all types. Here are some facts about cellular phone antennas that can help you to choose the best one for you.

FIGURE A.1
ANTENNAS

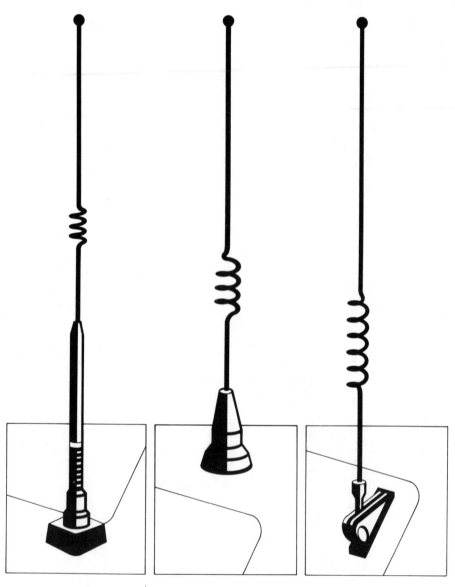

Trunk-mounted antenna.   Roof-mounted antenna.   Through-the-glass antenna.

ROOF-MOUNTED ANTENNAS

Ideally, the place for a mobile antenna is on the roof of
your car. First, this is the highest point on the vehicle so
the signals emitted from and coming into the antenna are
less likely to be obstructed by nearby objects. Second, to
function best, an antenna needs a *ground plane*, a surface
that actually works as a part of the antenna system to
"launch" radio waves, even though there is no electrical connection between it and the antenna. A ground plane should
have a radius equal to at least one-quarter of the wavelength of the radio wave. Since the frequencies used by cellular phones wavelengths are quite short—a little under a
foot—very little in the way of a ground plane is needed.
The roof of a car—provided it's made of metal, of course,
and not some material such as fiberglass that is transparent
to radio waves—provides more than enough surface area.

Roof-mounted antennas are not intended for trunk mounting. They are designed differently, and their performance
suffers if they are used in a situation for which they are not
intended. This holds true for both permanent and tempo-
rary antennas.

There are several reasons, however, why most antennas
are *not* roof mounted. The first is because it's inconvenient
to lead the antenna cable into the car from an antenna
mounted on top. It used to be that most cars had dome
lights in their ceilings, and these provided a means for con-
cealing the entrance of the cable (which came into the car
through the bottom of the antenna, which, itself, was
mounted in a hole cut in the roof). The cable could then be
routed inside the ceiling liner toward the trunk or front of
the car, where it had a relatively short exposed run to the
radio equipment.

Today's cars, if they have dome lights at all, usually have
them mounted toward the front of the roof, which means
that if an antenna were mounted over them, the ground
plane would be asymmetrical and probably not extend far
enough forward. For an antenna to radiate an omnidirectional signal—one that radiates out in all directions from

the antenna—it should be mounted as close to the center of the ground plane area as possible.

Furthermore, since a roof-mounted antenna extends at least a foot or so above the top of the car, it can easily be damaged by low garage doors and low-headroom passages such as the inside of a car wash. A roof-mounted antenna is also a permanent installation and cannot usually be easily removed and put back in place. (The magnetic-mount antennas discussed later are a different case.)

TRUNK-MOUNTED ANTENNAS

The next best location for an antenna is on a car's trunk lid since, after the roof, it offers the largest surface area for a ground plane. Although you sacrifice a little height by mounting an antenna there rather than on the roof, *trunk mounting* can offer benefits that outweigh the losses.

Trunk mounting is much easier and more convenient than putting an antenna on the roof. Putting a hole in the trunk of your car is not nearly as traumatic as putting one through the roof, and it's much simpler since not as much attention has to be paid to cosmetic detail. It is also possible to mount the antenna with a clip that attaches it to the edge of the trunk and to lead the antenna cable into the trunk through the space between the trunk and the body of the car. Some ground plane efficiency is sacrificed this way, but this method is still practical.

One antenna frequently used for trunk mounting is the *elevated-feed* antenna, where the signal is injected into the antenna at a point above its base. This serves two purposes. First, it makes the performance of the antenna resemble that of a roof-mounted one, and, second, it eliminates the need to rely on the car trunk to provide a ground plane.

For several reasons—appearance and security among them—a cellular phone's transceiver/logic unit is frequently mounted in the trunk of a car and connected to the rest of the system by cables. This makes it very convenient to connect the transceiver to a trunk-mounted antenna, which is right above it.

THROUGH-THE-GLASS ANTENNAS

One problem with a permanently mounted antenna of the type just described is that if you decide to remove it when you sell your car or transfer your phone, you have a hole that has to be filled. Probably the least expensive solution to this problem is to leave the old antenna where it is and buy a new one for the new installation. But an increasingly popular type of antenna—and one that requires no holes to be drilled—is the *through-the-glass* type, which is mounted on the car's rear window. While not quite as efficient as a permanently-mounted vehicle antenna, the *glass* mount performs satisfactorily under most conditions. Because it does not require a drilled hole and has a pleasing appearance, it has become the standard for vehicle installations.

While some owners like to display their glass mount antennas at a rakish angle, it is important that it be positioned perpendicular to the ground for optimum performance.

The antenna is in two parts. The "antenna" portion, which is attached to the outside of the window, consists of the antenna rod, base, and a mounting plate. The mounting plate is cemented to the window with a special weatherproof adhesive, and the base (which may also act as a mounting plate) contains a circuit that substitutes for the metal ground plane that would otherwise be provided by the roof or trunk lid of the car. The other part of the antenna is glued to the inside of the window, directly opposite the outside portion. The signal to and from the antenna is coupled through the glass by capacitive action, which allows a current to develop between the two sections of the antenna system without any physical connection between

 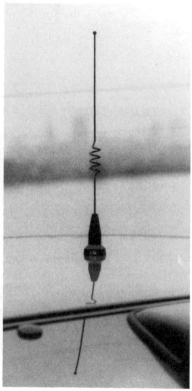

Through-the-glass antenna.   Roof-mounted antenna.

them. A coaxial cable that disappears into the car's rear deck connects the inside half of the antenna to the transceiver/logic unit, which, as usual, is probably located in the trunk.

Although through-the-glass antennas operate without ground planes, their efficiency can be affected by the presence of anything that acts as a ground plane, such as the demisting elements that may be a part of a car's rear window (and are required by law in some states). While the antenna will still work in the presence of these elements, it might not work as well.

The inside and the outside of the through-the-glass antenna must be aligned extremely carefully. If they are not, the efficiency of the antenna will be greatly reduced.

FLEX ANTENNAS

Flex, or "rubber duckie," antennas that look like stubby black breadsticks are frequently used with handheld and transportable phones. They are manufactured by packaging a wire and a flexible core inside a protective rubber or synthetic coating.

Flex antennas are not particularly efficient, but their small size and unobtrusiveness makes them popular with units that are to be carried. If you are using a transportable phone with a flex antenna in a marginal signal situation, try placing the phone (with the antenna) upright on a large metal object such as the roof of a car. The metal will act as a ground plane and improve the antenna's transmission and reception characteristics.

MAGNETIC-MOUNT ANTENNAS

Magnetic-mount, or mag-mount, antennas are intended for easy installation and removal on the roof or trunk lid of a vehicle. In its heavy base, a mag-mount antenna contains a strong permanent magnet that holds the unit firmly to the metal surface. The cable is usually let into the car, or into the trunk of the car if that's where the phone is located, through the doorframe or edge of the trunk lid. There is usually sufficient clearance to prevent it from being pinched, and the insulation around the door or trunk lid keeps water and wind from getting in.

If you use or store your car in an area where thefts are common, a removable mag-mount antenna can make your cellular phone installation less noticeable.

FEEDLINES

Radio signals are transported to and from an antenna by a type of feedline called coaxial cable, or "coax" for short. This cable consists of a center conductor surrounded by an electrical insulating layer and then by a shield of (usually) very thin braided copper wire. The whole is then encased in an outer layer of material to insulate it electrically and from the elements.

The most commonly used type of coax is designated RG-58, which is usually imprinted on the outer layer of insulation. RG-58 coax is similar to the RG-59 variety used in cable TV and master antenna installations.

Although this type of cable is thin (about one-quarter inch in diameter) and flexible, making it well suited for use in mobile installations, it has one major shortcoming. Like all coaxial cables using a braided shield, it is lossy at high frequencies—that is, the higher the frequency of operation (and the 800-MHz range of cellular phones is considered pretty high) the more radio frequency energy leaks out, meaning that less of it gets to (or from) the antenna. In long cable runs this loss can become significant.

Antenna coax is usually terminated in a plug-in connector with a screw-on shell to hold it in its socket. While most cellular phones and antennas use the same type of connector, if you do your own cellular installation make sure that the antenna connector on the transceiver and the plug on the end of the antenna cable are compatible.

THE PHASING COIL

The most visible characteristic of many cellular antennas, and one which sets them apart from most other types, is the little "pigtail" squiggle partway up its length. This is more than just decoration. It's called a *phasing coil* and serves to divide what looks like a single antenna into two antennas, one above it and the other below. Such an arrangement increases the efficiency of the antenna by about 50 percent.

The purpose of the phasing coil is to maintain the proper phase relationship of the radio waves so that each half of the antenna gets the portion of the signal intended for it.

Glossary

A-B switch   A switch on your cellular phone (which may be activated by pressing several keys on the keypad rather than being a physical device) that allows you to change between A (nonwireline) and B (wireline) carriers. A or B priority is usually automatic, with the other service being switched to only when the first is unavailable. Some phones allow you to lock them to an A or B carrier. *See also* Roaming, Nonwireline, Wireline

A carrier   *See* Nonwireline

Access charge   A flat monthly fee charged a subscriber for the use of a cellular system (whether he makes or receives any calls or not). Depending on the service plan chosen, this charge may include a number of "free" minutes of connect time. *See also* Connect time

Access number   The phone number that must be dialed by someone calling you when you are roaming, prior to dialing the number of your phone. The access number gives the caller access to the facilities of the system in which you are roaming. *See also* Roaming

AMPS   Advanced Mobile Phone Service. The official name for the cellular system. Also, an interconnection standard promulgated by AT&T for use in cellular phone equipment.

Antenna   A length of wire that radiates or captures radio signals.

Area code   The first three digits of your phone number (the middle one of which is always 0 or 1) that identify the part of the country in which it is serviced. The local calling area of your cellular service may include more than one area code. With cellular phones, it is not necessary to dial 1 before entering the area code. *See also* NPA

Band   A portion of the radio frequency spectrum. Cellular communications take place in the 800-MHz region of the UHF band. *See also* Block, UHF

B carrier   *See* Wireline

Block (of frequencies)   A group of radio frequencies within a band set aside for a particular purpose. Cellular telephony uses four blocks of frequencies within the 800-MHz portion of the UHF band. Nonwireline and wireline carriers are assigned separate blocks of frequencies, as are cell sites and mobile units. *See also* Band

Burn (a NAM)   To program information into a PROM. The device that does this is called a PROM burner or PROM programmer. *See also* NAM, PROM

Carrier   A company that provides telephone (or another communications) service. Also, an unmodulated radio signal. *See also* Nonwireline, Wireline

CDMA   Code Division Multiple Access. A digital cellular technology that tags digitized segments of a conversation with a unique code. This permits the segment to occupy a broad range of frequencies, to be identified and re-assembled into voice at the other end by having the correct code. This mechanism allows even greater channel capacity than TDMA. *See also* TDMA

Cell   The area assigned to a fixed-location cellular transmitting and receiving station. Each cell in a cellular system uses a pair of frequencies, to a maximum of forty-five, specific to it (within a region). The traffic from a mobile telephone is switched from cell to cell as the phone moves out of one and into another. *See also* Cell site, Handoff

Cell site   Where the fixed end of a cellular link is located. Also, the equipment used there. *See also* Cell

Cell splitting   Dividing one cell into two or more new cells to provide additional capacity within the original cell's region of coverage.

Cellular   Made up of cells; using cellular phone technology.

Channel   A frequency pair. There are currently 832 channels (including those for control signals) assigned to cellular use. *See also* Control signal, Frequency pair

Chip   *See* IC

CLEAR key   The key you press to erase information entered by mistake into a phone's memory. *See also* Memory

Coaxial cable (coax)   A type of cable used to carry signals in radio and other electronic communication. It consists of a center conductor surrounded by a cylindrical layer of insulation, which, in turn, is covered by a tube of braided copper wire or of thin foil. The whole is encased in a layer of protective insulation. The cable in cable TV is coaxial cable. *See also* Hard line

Connect time   The period your cellular phone is in radio contact with a cell site, not to be confused with the length of time your conversation lasts. Connect time is measured from the time your phone's IN USE indicator lights up until the time it goes off when you press the END key. *See also* END key, IN USE indicator, SEND key

Control head   The part of a cellular phone installation, generally located near the handset, that acts as the "go-between" between the phone user and the transceiver/logic unit. In some phones, the control head *is* the handset.

Control signal   A signal sent by a cell site to a cellular phone, or vice versa, carrying information necessary to the operation (and cooperation) of the two but *not* including the audio portion of a conversation. The channels used for control signals are separate from those used for voice. Control signals also flow between a cellular phone's handset and its transmitter/logic unit.

CPU   Central Processing Unit. *See also* Microprocessor

Dead spot   A location in a cellular system where, for one reason or another, signals do not penetrate. *See also* Multipath

Directional antenna   An antenna or array of antennas designed to concentrate a radio signal in a particular area. It also provides better reception of signals coming from the direction in which it is aimed than it does of signals coming from other directions. Directional antennas are frequently employed when a cell is split. *See also* Cell splitting, Omnidi-rectional antenna

Dropout   A momentary loss of signal. Brief dropouts dur-ing a cellular phone conversation can be annoying; longer ones can result in your being disconnected by the system.

DTMF   Dual-Tone Multi-Frequency. The official name for Touch Tone.

Dual-Mode   A cellular phone that can operate in stan-dard analog and the newer digital mode.

Dual NAM   A cellular phone feature that allows it to be registered on more than one cellular system. Multi-NAM phones are also available.

Duplex   Two-way. Cellular phones, using separate fre-quencies for transmission and reception, allow for duplex communications by allowing both parties to talk and listen at once. Push-to-talk systems are not duplex. *See also* Push-to-talk

END key   The key you press to terminate a call and dis-connect you from a cell site's equipment.

Error checking   A procedure performed by modems or by computer communications programs to make sure that information sent by modem gets from one computer to the other in good shape. If an error is detected, a request is usually made for the information in question to be re-transmitted.

ESN   Electronic Service Number. The serial number of a cellular phone. This information must be contained in the phone's NAM. *See also* NAM

FCC   Federal Communications Commission. The government agency responsible for regulating, among other things, cellular telephony.

Feedline   The cable between a radio and antenna that carries radio signals between them. Also called a "transmission line." *See also* Coax, Hard line

Fiber optics   The technology that allows electronic signals to be sent in the form of light through hairlike glass "pipes." This technology is being used increasingly in telephone and other communications systems.

Flex antenna   An antenna used with portable phones. It consists of a length of stiff wire, usually with a fiberglass core, covered with an insulating material. Sometimes called a "rubber duckie."

FM   Frequency Modulation. A mode of radio transmission particularly immune to interference. Cellular telephones use FM.

Full spectrum   These phones allow the user to access all 832 channels. Previously, phones could only access 666 channels.

Frequency pair   A set of two frequencies, referred to together as a "channel," used in cellular communications. One frequency is used for transmission, the other for reception. Used together they provide duplex communications. *See also* Channel, Duplex

Gelled-electrolyte battery   A type of battery made up of lead-acid cells whose acid electrolyte has been "jellied" to prevent spilling.

Ground plane   The surface of land or metal directly beneath an antenna that causes it to radiate and receive radio signals more effectively. A ground plane should be at least a quarter-wavelength in radius of the frequency for which it is intended. *See also* Wavelength

Handheld phone   A portable phone small enough to be carried and used in the hand. *See also* Portable

Handie-Talkie radio   A self-contained transceiver small enough to be carried and used in the hand. The radios police carry in holsters or in their hip pockets are Handie-Talkies, as are handheld cellular phones.

Handoff   The transfer of responsibility for a call from one cell site to the next. *See also* Cell site

Handset   The part of a phone you pick up and put to your ear. It contains a miniature speaker to reproduce sound and a microphone to pick it up. The handsets of cellular phones also usually contain a keypad for entering phone numbers and commands and a means of displaying those numbers and responses from the phone. *See also* Keypad, LCD

Hands-free phone   A phone that allows you to talk and listen via a microphone and speaker rather than through a handset you have to hold to your ear. Some hands-free phones can even recognize spoken words, allowing you to dial numbers while keeping your hands on the wheel.

Hard line   A special type of coaxial cable that uses a solid, rather than braided, outer shield. Hard line is frequently used in fixed cellular installations (cell sites) because of its high efficiency at UHF frequencies. *See also* Coaxial cable

HORN key   The key you press to turn on the "horn alert" circuit that beeps your car's horn when a call is received.

IC   Integrated Circuit. A quarter-inch "chip" of silicon on which can be contained the equivalent of thousands of transistors and other electronic components. The silicon chip is usually empackaged in plastic, with wire leads to be connected to other parts of an electronic circuit.

Immobilized-electrolyte battery   A type of battery made up of lead-acid cells whose acid electrolyte is contained by a spongelike material to prevent spilling.

IMTS   Improved Mobile Telephone Service. Cellular telephony's predecessor, which used a single central transmitter and receiver to service a region. *See also* Cell, Cellular

IN USE indicator   The indicator that tells you when your phone is in contact with a cell site. As long as it is lit, you are paying for the connect time on a per minute basis.

Keypad   A set of pushbutton electronic switches. The keys on a calculator make up a keypad, as do the buttons on a cellular phone.

kHz   Kilohertz. A measure of audio and radio frequency (a thousand cycles per second). The human ear can hear frequencies up to about 20 kHz. There are 1000 kHz in 1 MHz.

Landline   Traditional, wire-based telephone service, to distinguish it from cellular.

LCD   Liquid Crystal Display. The type of "readout" used by most cellular phones to display the digits of phone numbers as they are dialed or recalled, and other information.

Lead-acid cell   A type of rechargeable power cell using an acid electrolyte. Your car battery is made up of lead-acid cells. See also Gelled-electrolyte cell, Immobilized-electrolyte cell

LOCK key   The key you press to "lock" your phone against unauthorized use. The phone is unlocked by entering a personalized four- or five-digit code.

Logic unit   The computer section of a cellular phone, usually combined in the same package with the transceiver.

Magnetic-mount antenna   A type of antenna for mobile use. It has a magnetic base that sticks the antenna to a vehicle's roof or trunk lid and allows it to be removed quickly. A magnetic-mount antenna is a worthwhile investment if you use or store your car in a high-risk area, or if you use your phone in a number of different vehicles.

Memory   The ICs in a cellular phone that store phone numbers for instant recall. *See also* IC

MHz   Megahertz. A measure of radio frequency. One MHz is one million cycles per second. Cellular signals fall in the 800- to 900-MHz portion of the radio frequency spectrum. *See also* kHz, RF

Microprocessor   The IC that is the heart and brains of a small computer or computer-controlled device (such as a cellular phone). Frequently referred to as a CPU. *See also* IC

Millisecond   One-thousandth of a second.

Milliwatt   One-thousandth of a watt. Handheld cellular phones usually have a maximum output power of 600 milliwatts.

Mobile   Moving. Mobile phones are usually found in vehicles. Portable phones can be mobile, but mobile phones are not necessary portable. *See also* Portable

Modem   MOdulator/DEModulator. A device used to send information from one computer to another over a telephone system.

MTSO   Mobile Telephone Switching Office. Located between a cell site and a conventional telephone switching office, an MTSO is the link between a cellular phone and the rest of the phone system. An MTSO also handles the routing of traffic within a system.

Multipath   A condition where a signal from one source is received by several (a direct and any number of reflected) paths. This can frequently cause distortion or loss of signal in cellular communications.

NAM   Numeric Assignment Module. A type of integrated circuit called a PROM that is programmed to contain information specific to your cellular phone such as its ESN and the phone number assigned to it. The information contained in its NAM is what identifies your phone to a cell site and MTSO. *See also* ESN, IC, PROM

Nickel-cadmium cell   A type of "dry" rechargeable power cell. Sometimes called by its trade name, "Nicad."

Nonwireline   Refers to a cellular carrier that has no involvement in providing conventional telephone services. Nonwireline carriers are also known as A carriers. *See also* A-B switch, Carrier

NO SERVICE indicator   The indicator that tells you when you are in an area where no cellular service is available, or temporarily impossible.

NPA   Number Plan Area. Another name for your local cellular calling and billing area. This may include neighboring area codes in addition to the one assigned your phone.

Obsolete, obsolescent   Replaced by something newer. An obsolete or obsolescent cellular phone works just as well as the one that replaces it—it just lacks one or two of the latest convenience features.

Off-peak time The hours during which a system is least used. In cellular systems, off-peak time is usually defined as being between 7:00 p.m. and 7:00 a.m., as well as weekends and holidays. Rates are usually lowest during this period. *See also* Peak time

Omnidirectional antenna An antenna that is equally effective in all directions. *See also* Directional antenna

Operating life The length of time that a set of batteries will power a handheld or portable phone, usually four to eight hours. This period, as specified, is the maximum possible with the phone in "standby" mode—off the air waiting for a call. Actually using the phone decreases the operating life by a factor of about six (one hour of use equals about six hours of standby).

PCS Personal Communications Services. A new family of wireless services yet to be fully defined, which may include new radio frequencies, cellular-like services that locate users anywhere at one telephone number, and new data and messaging services.

Peak time The hours of heaviest usage of a system. In cellular systems, peak time is usually defined as being between 7:00 a.m. and 7:00 p.m., Monday through Friday. Usage rates are higher during peak time than during off-peak time. *See also* Off-peak time

Phasing coil The "pigtail" in a cellular phone antenna that divides the two sections of the antenna and splits a radio signal between them.

Portable Capable of being carried from place to place. Portable cellular phones contain their own power supplies and can be used anywhere there is service. Portables are further distinguished from transportables by their smaller size and lower power (600 milliwatts instead of 3 watts).

Program   To store information for carrying out instructions in a computerized system. A NAM is programmed with such information as a phone's ESN and phone number, which allow it to be recognized by a cellular system. *See also* Burn, ESN, NAM, PROM

Push to talk   A type of radio communication requiring a speaker to push a button or lever to activate the transmitter. When released, the transmitter turns off, and the receiver comes into action. You can talk or listen with this type of arrangement but not both at the same time. *See also* Duplex

Radiotelephone   A mobile or portable telephone that uses radio to link it to the rest of the nation's phone system. Cellular phones are radiotelephones, although they are not usually referred to as such.

RCL (ReCalL) key   The key you press, along with a couple of digits, to retrieve a number from your phone's memory. *See also* Memory

Reuse (of frequencies)   Assigning frequencies to cell sites so that no adjoining cell sites use the same ones. Cell sites out of range of one another can use (or reuse) the same frequencies.

RF   Radio Frequency. Refers to electromagnetic waves having frequencies between about 10 kHz and 300,000 MHz. Sometimes also used to refer to the radio signal itself. *See also* MHz

ROAM indicator   The indicator that tells you when you are out of your local service area and in that of another cellular carrier. If the carrier in the "foreign" area is different from the one the phone is set up for (wireline instead of nonwireline, or vice versa) the ROAM indicator may flash to indicate this. *See also* A-B switch

Roaming   Using your cellular phone outside of its "home" service area. Some cellular systems have automatic roaming agreements with others; other systems require you to register with them before you can use their facilities to place or receive calls.

Rubber duckie   *See* Flex antenna

SEND key   The key you press to put your cellular phone "on the air."

Service area   The region covered by the cellular service to which you subscribe. Also known as your "number plan area." *See also* NPA

Smarts   The "intelligence" imparted to a device by the computer built into it.

Subscriber   One who pays for service from a system.

STO(re) key   The key you press to cause information to be stored in a phone's memory. *See also* Memory

TDMA   Time Division Multiple Access. A digital cellular technology that assigns digitized segments of a conversation to one of several time slots on a single channel. *See also* CDMA

Traffic   The communications carried by a system.

Transceiver   A radio transmitter and receiver combined to form a single unit. A cellular phone uses a transceiver to send signals to, and receive them from, the cell site.

Transportable   *See* Portable

UHF   Ultra High Frequency. The UHF band, in which fall the frequencies used by cellular phones, ranges from 300 MHz to 3000 MHz. *See also* MHz

Voice messaging   The ability to receive, store, and send voice messages without being on the phone at the same time with another. It allows cellular users to be notified of incoming calls and take messages when their cellular phone is out of the area or turned off, in combination with call forwarding.

Wavelength   The distance from a point on a radio wave to the corresponding point on the next wave. The higher the frequency of a radio signal, the shorter its wavelength. The signals used in cellular telephony have a wavelength of about twelve inches.

Wireless   The set of radio-based communications services, including cellular, paging, other mobile radio services, and PCS.

Wireless phone   A phone, used in conjunction with another phone, that is not directly connected to the phone line. Although cellular phones are wireless, they are not referred to as wireless phones.

Wireline   Refers to a cellular carrier that is also involved in providing conventional telephone service. Wireline carriers are also known as B carriers. *See also* A-B switch, Carrier

Index

Photo Credits

We would like to thank the following for the use of their photos:

page 12 Nokia Mobile Phones
page 20 Quantum Publishing
page 33 Motorola, Inc.
page 70 RACOM
page 71 Audiovox Corp
page 72 Nokia Mobile Phones
page 74 NEC (drawing based on NEC photo)
pages 75, 86 Walker Telecommunications
page 77 OKI Telecom
page 87 Interstate Voice Products
page 85 ORA Electronics
page 95 Spectrum Cellular
page 134 ORA Electronics
All other photos courtesy of Radio Shack.